OPERATOR COLLIGATIONS IN HILBERT SPACES

OPERATOR COLLIGATIONS IN HILBERT SPACES

Mikhail S. Livshits and Artem A. Yantsevich

Translated by the
American Mathematical Society

Edited and introduced by
Ronald G. Douglas
SUNY at Stony Brook

1979

V. H. WINSTON & SONS
Washington, D.C.

A HALSTED PRESS BOOK

JOHN WILEY & SONS
New York Toronto London Sydney

V. H. Winston & Sons, a Division of Scripta Technica, Inc., Publishers
1511 K Street, N.W., Washington, D.C. 20005

Distributed solely by Halsted Press, a Division of John Wiley & Sons, Inc.

Library of Congress Cataloging in Publication Data:

Livshits, Mikhail Samuilovich.
Operator colligations in Hilbert spaces.

Bibliography: p.
1. Operator theory. 2. Hilbert space.
3. Stochastic processes. I. Yantsevich, A. A., joint author. II. Douglas, Ronald G.
III. Title. IV. Series.
QA329.L58 515'.72 78-13073
ISBN 0-470-26541-8

CONTENTS

FOREWORD

The evolution of physical systems under discrete time can often be described by the action of an operator on some linear space where the norm is taken to be energy. And if energy is given by a quadratic form, one has a space which when completed yields a Hilbert space. Moreover, if energy is conserved then the operator in question is unitary and hence the exponential of a self-adjoint operator. From this arises the importance of self-adjoint operators in most of the early applications of operator theory to the study of physical systems.

In not all systems, however, is energy conserved. Generally systems are dissipative and hence one is led to study dissipative operators. Such a study was begun by M. S. Livshits in the late forties and he introduced the notion of characteristic operator function. This has been intensively developed during the last thirty years by many authors and has had far reaching applications.

In the first part of this book the authors extend and enlarge upon this by further developing the notion of an operator colligation which arose out of the earlier work. Contact is made with Riemannian geometry and applications to what the authors call open systems are hinted at.

In the second part of the book the authors tackle the subject of nonstationary stochastic processes. During the last forty years the prediction theory of stationary stochastic processes has been systematically developed with spectacular success. Such processes are described by a unitary operator. Again processes coming from the real world are likely not to be stationary and the authors show how the earlier work on dissipative operators can be utilized to study such processes. Lastly, operator colligations which are invariant under a transformation group are studied.

We conclude by observing that just as in the pioneering work of thirty years ago, the mathematics is interesting, deep, and will more than likely prove appropriate for application.

Ronald G. Douglas

PREFACE

Commencing in the fifties, the development of the theory of non-self-adjoint operators has led to the creation of spectral analysis and the theory of elementary divisors for various classes of linear operators in Hilbert space. The further development of this theory, which has found applications to various questions in mathematics and physics (we mention in this connection the monographs of M.S. Brodskii [4], I.C. Gohberg and M.G. Kreĭn [16^1, 16^2], M.S. Livshits [27^1] and B. von Sz.-Nagy and C. Foiaş [40] as well as the works of H. Helson [17] and L. de Branges and J. Rovnyak [3], where closely related questions are considered), continues to the present day.

As a result of the evolution of the notion of a characteristic function of an operator it has become apparent that in many questions the natural object of investigation is not the operator itself but a more complicated object, known as an *operator colligation* or simply a *colligation*. The space in which an operator is defined is connected in an operator colligation to another space by means of a mapping whose role is to transfer a metric. The deep connection between operator colligations and open physical systems lies in the fact that if the motion of the system involves a change in its energy, there must exist a space playing the role of a window through which an interaction with the external world takes place. The value of a colligation is to be found in the fact that it explicitly contains not only the operator defining the motion but also a corresponding "window."

In Chapters I–III of the present monograph a calculus of colligations and of the associated open systems is developed and a connection is established with Riemannian geometry and tensor colligations.

In Chapter IV a study is made of the characteristic functions of a system of commuting operators.

In Chapter V and in §4 of Chapter IV the necessary information from the theory of characteristic functions and triangular models is presented, frequently without proof. The reader can find the corresponding proofs and a more detailed presentation of the questions touched upon here in the monograph of M.S. Brodskiĭ [4] and in the survey article of M.S. Brodskii and M.S. Livshits [5].

Chapters VI–VIII are devoted to applications to the spectral theory of nonstationary probability processes and their correlation functions. The *infinitesimal correlation function*

$$w(t, s) = -\frac{\partial}{\partial \tau} v(t+\tau, s+\tau)\Big|_{\tau=0},$$

where $v(t, s)$ is a correlation function, is used to introduce the notion of *nonstationariness rank*, which is defined as the maximal rank r $(0 \leqslant r \leqslant \infty)$ of all quadratic forms of the form

$$\sum_{j,k=1}^{n} w(t_j, t_k)\xi_j\bar{\xi}_k \quad (-\infty < t_1 < t_2 < \cdots < t_n < \infty, \ n = 1, 2, \ldots).$$

The form of the correlation function is found and spectral resolutions are obtained for various classes of linearly representable dissipative processes *of given rank r*. In contrast to the well-known Loève-Karhunen canonical expansion, an expansion in a chain of interconnected damped oscillators, the amplitudes of which are uncorrelated and the "eigenfrequencies" of which are complex, is obtained with the use of non-self-adjoint operators.

In Chapters IX–X, which were written by V.K. Dubovoĭ, a study is made of operator colligations that are invariant with respect to transformation groups and of open fields that are invariant with respect to the group of Lorentz transformations.

The reader should be acquainted with the theory of linear operators as presented in the first six chapters of the book of N.I. Ahiezer and I.M. Glazman [1].

We wish to thank V.E. Kacnel'son for his considerable help in the editing of the manuscript [5].

Mikhail Livshits and *Artem Yantsevich*

CHAPTER I

INDUCED METRICS AND OPERATOR COLLIGATIONS

§1. POLYMETRIC SPACES AND INDUCTORS

1. Let E be a Hilbert space ($\dim E < \infty$) and let $\mu(u, v)$ $(u, v \in E)$ be a Hermitian form on E, i.e. a functional $\mu(u, v)$ on $E \times E$ that is linear and continuous in u for each v and satisfies the identity $\mu(u, v) = \overline{\mu(v, u)}$. The form $\mu(u, v)$ can be used to introduce a metric in E by putting the square of the length of an element u equal to the quantity $\mu(u, u)$. This metric can be indefinite and degenerate. In what follows, any Hermitian form will be called a metric. Clearly, if μ_k $(k = 1, 2, \dots, n)$ are metrics defined on E, the set of all possible linear combinations $\sum_{k=1}^{n} c_k \mu_k$ of them with real coefficients c_k is a linear space of metrics on E.

Definition. A Hilbert space E is called *polymetric* if a set K_E of metrics is defined on it.

Definition. By the *radical* of a metric $\mu(u, v)$ defined on E is meant the set of elements u_0 such that $\mu(u_0, v) = 0$ for all $v \in E$. A set K_E of metrics is said to be *nondegenerate* if the intersection $\bigcap_{\mu \in K} \operatorname{Rad} \mu$ of the radicals of the members of K_E is empty. In this case the polymetric space E will also be called nondegenerate.

2. Let E and H be a pair of Hilbert spaces and let Φ be a continuous linear mapping of H into E $(H \overset{\Phi}{\to} E)$.

If E is a polymetric space while H is not, the mapping Φ can be used to transfer the set K_E of metrics given on E to a corresponding set K_H of metrics on H by putting

$$\mu_H(h_1, h_2) = \mu_E(\Phi(h_1), \Phi(h_2)) \quad (h_1, h_2 \in H).$$

These metrics will be called *induced metrics* and denoted by the symbol $\mu_H = \operatorname{Ind}\mu_E$ $(\mu_E \in K_E)$.

Definition. If E is a polymetric space, the quadruplet (H, Φ, E, K_E) is called a *metric inductor* or simply an *inductor*, while H and E are respectively called the *internal* and *external spaces* of the inductor.

An inductor is said to be *nondegenerate* if E is a nondegenerate polymetric space.

Definition. We will say that a Hilbert space H with a preassigned set K_H of metrics on it is *included in an inductor* (H, Φ, E, K_E) *if* $K_H = \operatorname{Ind} K_E$.

Theorem 1.1. *A Hilbert space* H *with a preassigned set* K_H *of metrics on it can always be included in a nondegenerate inductor* (H, Φ, E, K_E) *in such a way that* $\Phi(H) = E$. *If* K_H *is a linear space of metrics,* K_E *is also a linear space and* $\dim K_E = \dim K_H$.

Proof. Let $H_0 = \bigcap_{m \in K_H} \operatorname{Rad} m$, and let $H_1 = H \ominus H_0$ be the orthogonal complement of H_0 in the sense of the Hilbert metric. Consider the inductor (H, Φ, E, K_E), where $E = H_1$, the operator Φ is the orthogonal projection of H onto H_1 and the metrics $\mu \in K_E$ are the restrictions to H_1 of the metrics $m \in K_H$.

Since

$$m(h, g) = m(\Phi(h), \Phi(g)) = \mu(\Phi(h), \Phi(g)) \quad (h, g \in H),$$

$m = \operatorname{Ind}\mu$ $(m \in K_H)$.

The set K_E of metrics is nondegenerate. For suppose $u_0 \in \operatorname{Rad}\mu$ for $\mu \in K_E$. From the identity $\mu(u_0, v) \equiv 0$ for all $v \in E$ it follows that $m(u_0, g) \equiv 0$ for all $g \in H$ and hence $u_0 \in \operatorname{Rad} m$ for $m \in K_H$, i.e. $u_0 \in H_0 \cap E = 0$.

Since the identity $\mu(u, v) \equiv 0$ implies the identity $m(h, g) \equiv 0$, in the case when K_H is a linear space of metrics

$$\dim K_E = \dim K_H.$$

3. Consider an inductor $L = (H, \Phi, E, K_E)$. Fix $\mu \in K_E$. Clearly $\mu(\Phi(h), v)$ is, for fixed v, a continuous linear functional on H, and hence by Riesz' theorem

$$\mu(\Phi(h), v) = (h, \Phi^+(v)), \tag{1.1}$$

where the right side is the scalar product in H of the element $h \in H$ and some element $\Phi^+(v) \in H$.

The mapping $\Phi^+ (E \xrightarrow{\Phi^+} H)$ is linear, single-valued and continuous (in the Hilbert metrics of E and H); it depends on the choice of the metric $\mu \in K_E$ and will be denoted by Φ_μ^+.

Since an induced metric $m_\mu(h, g)$ satisfies the relations

$$m_\mu(h, g) = \mu(\Phi(h), \Phi(g)) = (\Phi_\mu^+ \Phi h, g),$$

we have

$$\operatorname{Ind} \mu = m_\mu(h, g) = (\Phi_\mu^+ \Phi h, g). \tag{1.2}$$

Thus a metric induced on H can be obtained from the operator $\Phi_\mu^+ \Phi$ acting from H into H by applying formula (1.2).

We note that if an inductor (H, Φ, E, K_E) is nondegenerate, the mapping Φ is uniquely determined by the mappings Φ_μ^+ defined for $\mu \in K_E$. For suppose $\Phi_{1\mu}^+ = \Phi_{2\mu}^+$. Then

$$\mu(\Phi_1(h) - \Phi_2(h), v) = (h, \Phi_{1\mu}^+(v) - \Phi_{2\mu}^+(v)) = 0$$

and consequently

$$\Phi_1(h) - \Phi_2(h) \in \bigcap_{\mu \in K_E} \operatorname{Rad} \mu.$$

§ 2. MONOMETRIC INDUCTORS

The special case of inductors in which the set K_E of metrics consists of a single element plays an important role in the sequel. Such inductors will be called *monometric*.

Two monometric inductors (H', Φ', E', μ') and $(H'', \Phi'', E'', \mu'')$ are said to be *H-equivalent* if $H' = H'' = H$ and $\operatorname{Ind}\mu' = \operatorname{Ind}\mu''$.

Theorem 1.2. *Any monometric inductor* (H, Φ, E, μ) *is H-equivalent to an inductor* (H, Φ', E, μ'), *where* $\mu' = (Ju, v)$ $(u, v \in E)$ *and the operator* J *is an involution in* $(J = J^*, J^2 = I)$.[1]

Proof. Since the metric μ is a Hermitian form on E, it has the representation $\mu = (Bu, v)$, where B is a *Hermitian operator*, i.e., a bounded self-adjoint operator.

We represent the operator B in the form $B = \int_a^b \lambda \, dE_\lambda$, where E_λ is a resolution of the identity, and put $K = \int_a^b |\lambda|^{1/2} dE_\lambda$ and $J = \int_a^b \operatorname{sign}\lambda \, dE_\lambda$. Then $B = KJK$, with $J = J^*$ and $J^2 = I$. (If $\dim E < \infty$, the spectral resolution of B need not be used.)

The metric $m(h, g) = \operatorname{Ind}\mu$ can be written as follows:

$$m(h, g) = \mu(\Phi(h), \Phi(g)) = (KJK\Phi(h), \Phi(g)) =$$
$$(JK\Phi(h), K\Phi(g)) = (J\Phi_1(h), \Phi_1(g)),$$

where $\Phi_1(h) = K\Phi(h)$.

The inductor $(H, \Phi_1, E, \mu = (Ju, v))$ clearly satisfies the conditions of the theorem.

A metric $\mu_J = (Ju, v)$ defined with the use of an involution J will be called a *J-metric*.

We note that in the case of a J-metric

$$\Phi_\mu^+ = \Phi^* J. \tag{1.3}$$

We consider some elementary operations taking a given inductor into inductors that are H equivalent to it.

[1] The notation (,) stands for the scalar product in E.

1) The passage from an inductor $L = (H, \Phi, E, \mu)$ to an inductor $L_1 = (H, \Phi, E \dotplus E_1, \mu \dotplus \mu_1)$ where E_1 is an arbitrary monometric space, will be called a *trivial lengthening* (the passage from L_1 to L will be correspondingly called a *trivial shortening*).

Here $E \dotplus E_1$ is the direct sum of two monometric spaces, the metric on $E \dotplus E_1$ being defined as follows:

$$\mu(\{u, u_1\}, \{v, v_1\}) = \mu(u, v) + \mu_1(u_1, v_1) \quad (u, v \in E, u_1, v_1 \in E_1)$$

2) The passage from an inductor $L = (H, \Phi, E, \mu)$ to an inductor $L_2 = (H, \Phi \dotplus \Phi_2 \dotplus \Phi_2, E \dotplus E_2 \dotplus E_2, \mu \dotplus \mu_2 \dotplus \mu_2^-)$, where E_2 is an arbitrary monometric space with metric μ_2 and $\mu_2^- = -\mu_2$, will be called a *neutral lengthening* (correspondingly, the passage from L_2 to L will be called a *neutral shortening*).

3) The passage from an inductor $L = (H, \Phi, E, \mu)$ to an inductor $L_3 = (H, \Phi_3, E_3, \mu_3)$, where $\Phi_3 = U\Phi$, $E_3 = UE$, $\mu_3(u_3, v_3) = \mu(U^{-1}u_3, U^{-1}v_3)$ and U is a unitary (in the sense of the Hilbert metrics) operator mapping E onto E_3, will be called a *unitary operation*.

We have the following

Theorem 1.3. *If L_1 and L_2 are two H-equivalent inductors with J-metrics $\mu_1 = (J_1u_1, v_1)$ and $\mu_2 = (J_2u_2, v)$, one of them can be obtained from the other by means of a finite number of unitary, lengthening and shortening operations.*

Proof. *We lengthen the inductor L_1 as follows:*

$$(H, \Phi_1 \dotplus \Phi_2 \dotplus \Phi_2, E_1 \dotplus E_2 \dotplus E_2, \mu_1 \dotplus \mu_2 \dotplus \mu_2^-)$$

The theorem will be proved if the inductor $L_3 = (H, \Phi, E, \mu)$, where $\Phi = \Phi_1 \dotplus \Phi_2$, $E = E_1 \dotplus E_2$ and $\mu = \mu_1 \dotplus \mu_2^-$, is reduced to the trivial form $(H, 0\ 0\ 0)$ by means of the above mentioned elementary operations. Clearly, the metric $\mu(u, v)$ has the form $\mu(u, v) = (Ju, v)$, where $J = J_1 \dotplus J_2^-$ is an involution with $J_2^- = -J$.

Since $\operatorname{Ind}\mu = \operatorname{Ind}\mu_1 + \operatorname{Ind}\mu_2^-$, we have

$$\mu(\Phi h, \Phi h) = 0 \quad (h \in H) \tag{1.4}$$

The inductor L_3 can be represented in the form $L_3 = (H, \Phi_+ \dot{+} \Phi_-, E_+ \dot{+} E_-, \mu_+ \dot{+} \mu_-)$, where $E_\pm$ are the eigenspaces of J belonging respectively to the eigenvalues ± 1, the operators $\Phi_\pm = P_{E_\pm}\Phi$, where $P_{E_\pm}$ are the orthogonal projections of H onto $E_\pm$ and $\mu_\pm(u, v) = (u, v)$ $(u, v \in E_\pm)$.

The subspaces $E_\pm$ can be represented in the form $E_\pm = E'_\pm \oplus E^\circ_\pm$, where $E'_\pm = \overline{\Phi_\pm H}$ are the closures of the ranges of the operators $\Phi_\pm$.

As the result of a trivial shortening of the inductor L_3 we obtain the inductor $L_4 = (H, \Phi'_+ \dot{+} \Phi'_-, E'_+ \dot{+} E'_-, \mu'_+ \dot{+} \mu'_-)$, where $\mu'_\pm$ are the restrictions of the metrics $\mu_\pm$ respectively to the subspaces $E'_\pm$ and $\Phi'_\pm = \Phi_\pm$.

By virtue of relation (1.4) we have

$$(\Phi'_+ h, \Phi'_+ h) - (\Phi'_- h, \Phi'_- h) = 0$$

Therefore the operator U' taking the vector $\Phi'_- h$ into the vector $\Phi'_+ h$ maps isometrically the range of Φ'_- onto the range of Φ'_+. Extending the operator U' onto all of E'_- by continuity, we obtain a unitary operator U'. Thus Φ'_+ admits a representation of the form

$$\Phi'_+ = U\Phi'_- \tag{1.5}$$

where U' is a unitary operator mapping E'_- onto E'_+.

Substituting (1.5) into L_4, we obtain the inductor

$$(H, U'\Phi'_- \dot{+} \Phi'_-, E'_+ \dot{+} E'_-, \mu'_+ \dot{+} \mu'_-)$$

From this inductor, by means of an obvious unitary operation and a neutral shortening, we arrive at the trivial inductor. The theorem is proved.

§3. OPERATOR COLLIGATIONS

Let A be bounded linear operator defined in H and let $L = (H, \Phi, E, \mu)$ be a monometric inductor.

Definition. The pair $M = (A, L)$ is called an *aggregate*, while the operator A is called an *internal operator* of the aggregate M.

Definition. An aggregate (A, H, Φ, E, μ) is called a *metric colligation* if

$$((AA^* - I)h, g) = \operatorname{Ind}\mu. \tag{1.6}$$

And if an aggregate satisfies the condition

$$((A + A^*)h, g) = \operatorname{Ind}\mu, \tag{1.7}$$

it is called a *local colligation*.

Relations (1 . 6) and (1 . 7) mean the agreement of certain metrics generated by the operator A in H with the induced metric.

Conditions (1 . 6) and (1 . 7) can be rewritten with the use of (1 . 2) in the forms

$$AA^* - I = \Phi^+\Phi, \tag{1.8}$$

$$A + A^* = \Phi^+\Phi. \tag{1.9}$$

Thus metric and local colligations can be defined by commutative diagrams (Figures 1 and 2).

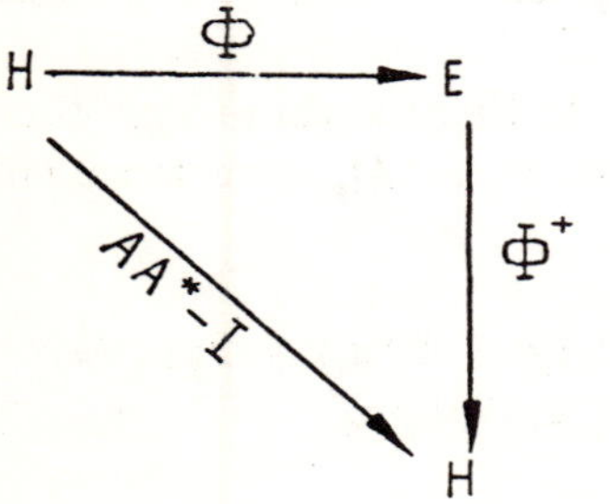

Fig. 1.

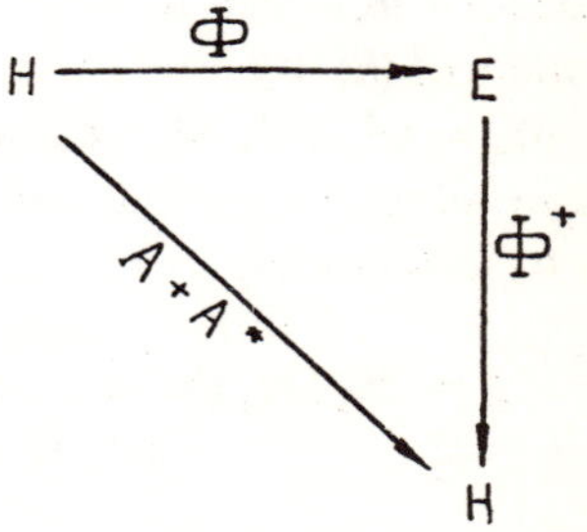

Fig. 2.

In the sequel we will encounter more general constructions of the form $\vec{X} = (A_1, A_2, \ldots, A_n, H, \Phi, E, \mu_1, \mu_2, \ldots, \mu_n)$ with n internal operators and n metrics in the space E. Such constructions will be called *vector aggregates* .

Finally, if

$$A_k + A_k^* = \operatorname{Ind} \mu_k = \Phi_k^+ \Phi \quad (k = 1, 2, \ldots, n)$$

a vector aggregate is called a *vector local colligation* or simply a *local colligation.*[2]

We note that if A is a unitary operator in H, it can be included in a metric colligation with the trivial inductor; in addition, there exists an uncountable set of metric colligations with nontrivial inductors in which a unitary operator can be included . These colligations can be obtained, in particular, by carrying out neutral lengthenings of the trivial inductor.

Suppose we are given an aggregate (colligation) $M = (A, H, \Phi, E, \mu)$. For each fixed element $u \in E$ we can associate with this aggregate a mapping $g = M(h;\ u)$ of H into H defined by the equality

$$g = Ah + \Phi^+ u \quad (h \in H,\ u \in E) \tag{1.10}$$

If the element u ranges over the whole Hilbert space E then $M(h;\ u)$ $(h \in H)$ maps $H \dotplus E$ into H.

In what follows, each mapping of the form (1.10) will be called a *motion with external influence* u, while M will be called the aggregate (colligation) of the motion $M(h;\ u)$.

Let $M_k = (A_k, H, \Phi_k, E_k, \mu_k)$ $(k = 1, 2)$ be a pair of aggregates. Consider two successive motions with aggregates M_1 and M_2 and external influences u_1 and u_2 respectively:

$$\begin{gathered} g = M_2[M_1(h;\ u_1);\ u_2] = A_2(A_1 h + \Phi_1^+ u_1) + \Phi_2^+ u_2 = \\ (A_2 A_1) h + A_2 \Phi_1^+ u_1 + \Phi_2^+ u_2. \end{gathered} \tag{1.11}$$

[2] Vector metric operator colligations are not encountered in the sequel.

It is easily seen that equality (1.11) defines a new motion of the form $g = M(h;\ u_1 + u_2)$, where

$$M = (A_2A_1,\ H,\ \Phi_1A_2^* \dotplus \Phi_2,\ E_1 \dotplus E_2,\ \mu_1 \dotplus \mu_2). \tag{1.12}$$

The aggregate M defined by equality (1.12) is naturally called the *product* of the aggregates M_1 and M_2, and one writes $M = M_2M_1$. It can be directly verified that this product has the associativity property $M_3(M_2M_1) = (M_3M_2)M_1$.

We note that the result of two successive motions $M_1(h;\ u_1)$ and $M_2(h;\ u_2)$ is a motion with aggregate $M = M_2M_1$ and external influence $u_1 + u_2$.

The following assertion holds when the aggregates of the motions are metric colligations.

A product of two metric colligations is a metric colligation.

For suppose $M_k = (A_k,\ H,\ \Phi_k,\ E_k,\ \mu_k)$ $(k = 1,\ 2)$ are a pair of metric colligations, so that by virtue of (1.8)

$$A_kA_k^* - I = \Phi_k^+\Phi_k \tag{1.13}$$

Then for the aggregate $M = M_2M_1 = (A,\ H,\ \Phi,\ E,\ \mu)$, where $A = A_2A_1$, $\Phi = \Phi_1A_2^* \dotplus \Phi_2$, $E = E_1 \dotplus E_2$ and $\mu = \mu_1 \dotplus \mu_2$, we get, using (1.13),

$$AA^* - I = A_2A_1A_1^*A_2^* - I = A_2(I + \Phi_1^+\Phi_1)A_2^* - I =$$
$$A_2\Phi_1^+\Phi_1A_2^* + A_2A_2^* - I = A_2\Phi_1^+\Phi_1A_2^* + \Phi_2^+\Phi_2.$$

On the other hand,

$$\Phi^+\Phi = (A_2\Phi_1^+ \dotplus \Phi_2^+)(\Phi_1A_2^* \dotplus \Phi_2) = A_2\Phi_1^+\Phi_1A_2^* + \Phi_2^+\Phi_2$$

and hence $AA^* - I = \Phi^+\Phi$.

§4. PRODUCT OF LOCAL COLLIGATIONS

Let $X_k = (A_k,\ H_k,\ \varphi_k,\ E,\ \mu)$ $(k = 1,\ 2)$ be a pair of aggregates with a common external space E and a common metric μ.

Definition. An aggregate $X = (A, H, \varphi, E, \mu)$ is called the *E-product* of the aggregates X_1 and X_2 $(X = X_1 \vee X_2)$ if the following conditions are satisfied:

$$\begin{aligned} &1)\ A = A_1P_1 + A_2P_2 + \varphi_2^+\varphi_1P_1, \quad H = H_1 \oplus H_2, \\ &2)\ \varphi = \varphi_1P_1 + \varphi_2P_2, \end{aligned} \tag{1.14}$$

where the P_k are the orthogonal projections of H onto the H_k respectively.

Clearly, H_2 is an invariant subspace of A. An E product has the associativity property $(X_1 \vee X_2) \vee X_3 = X_1 \vee (X_2 \vee X_3)$.

Theorem 1.4. *An E product of local colligations is a local colligation.*

The proof is obtained by means of an elementary calculation. If $h = h_1 + h_2$ and $g = g_1 + g_2$ $(h_k, g_k \in H_k)$, then

$$\begin{aligned} ((A + A^*)h, g) &= (Ah, g) + (h, Ag) = (A_1h_1, g_1) + (A_2h_2, g_2) + \\ &(\varphi_2^+\varphi_1h_1, g_2) + (h_1, A_1g_1) + (h_2, A_2g_2) + (h_2, \varphi_2^+\varphi_1g_1) = \\ &((A_1 + A_1^*)\, h_1, g_1) + ((A_2 + A_2^*)\, h_2, g_2) + \mu\,(\varphi_1h_1, \varphi_2g_2) + \\ &\mu\,(\varphi_2h_2, \varphi_1g_1) = \mu\,(\varphi_1h_1, \varphi_1g_1) + \mu\,(\varphi_2h_2, \varphi_2g_2) + \mu\,(\varphi_1h_1, \varphi_2g_2) + \\ &\mu\,(\varphi_2h_2, \varphi_1g_1) = \mu\,(\varphi_1h_1 + \varphi_2h_2, \varphi_1g_1 + \varphi_2g_2) = \mu\,(\varphi h, \varphi g). \end{aligned} \tag{1.15}$$

Definition. By the *projection* P_0X (where P_0 is an orthogonal projection) of an arbitrary aggregate $X=(A, H, \varphi, E, \mu)$ onto a subspace H_0 $(=P_0H)$ is meant the aggregate $X_0 = (A_0 = P_0A, H_0, \varphi_0 = \varphi, \mu)$.

It is easily seen that the projection of a local colligation onto an arbitrary subspace is a local colligation.

Theorem 1.5. *If its internal space H has the form $H = H_1 \oplus H_2$ and H_2 is an invariant subspace of its internal operator A, a local colligation X is an E-product of its projections onto H_1 and H_2: $X = (P_1X) \vee (P_2X)$.*

Proof. The equalities $P_1AP_2 = 0$ and $(P_1 + P_2)\, A(P_1+P_2) = P_1AP_1 + P_2AP_2 + P_1AP_2 + P_2AP_1$ imply the following representation of A:

$$A = A_1P_1 + A_2P_2 + P_2AP_1. \tag{1.16}$$

Since $P_2A^*P_1 = (P_1AP_2)^* = 0$, the operator A has the form

$$A = A_1P_1 + A_2P_2 + P_2(A + A^*)P_1 = A_1P_1 + A_2P_2 + P_2\varphi + \varphi P_1 = A_1P_1 + A_2P_2 + \varphi_2^+\varphi_1P_1,$$

Q.E.D.

An immediate consequence of Theorem 1.5 is the following theorem on the decomposition of a local colligation.

Theorem 1.6. *If $H = H_0 \supset H_1 \supset H_2 \supset \cdots \supset H_{n-1} \supset H_n = 0$ is a decreasing chain of invariant subspaces of its internal operator A, a local colligation X admits a representation of the form*

$$X = X_1^\perp \vee X_2^\perp \vee \ \ldots \ \vee X_n^\perp, \tag{1.17}$$

where $X_k^\perp$ is the projection of X onto $H_{k-1} \ominus H_k$ $(k = 1, 2, \ldots, n)$.

NOTES

A local operator colligation was originally defined [4], [27¹] as a quintuple $X = (H, A, \Gamma, E, J)$ in which Γ is a linear mapping of E *into* H such that

$$(A - A^*)/i = \Gamma J \Gamma^*. \tag{1}$$

If the spaces are considered over the field of complex numbers, the left side of condition (1.9), i.e. twice the real part of A, can easily be replaced by twice the imaginary part.

Here we adopt the definition of a colligation proposed by E.M. Žmud' [43], who distinguished the metric inductor (called a dyad by him) as an independent notion and studied Lie groups and Lie algebras of colligations over an arbitrary field (of characteristic $\neq 2$), connecting them with the classical Witt groups [25].

The notion of a product of metric colligations was introduced in connection with the theory of nonunitary representations of groups [27²].

Theorem 1.3 was obtained by Do Hong Tan [12].

CHAPTER II

DIFFERENTIATION AND INTEGRATION OF COLLIGATIONS

§1. DIFFERENTIATION OF INDUCTORS

Let R_n be an n-dimensional space in which a system of coordinates is given, i.e. in which each point $Q \in R_n$ is determined by n coordinates $Q(x^1, x^2, \ldots, x^n)$. With every piecewise smooth arc curve $\gamma \in R_n$ we associate a space $L_2(\gamma, E)$ of functions $u(Q)$ $(Q \in \gamma)$ whose values belong to a given Hilbert space E.

A scalar product is defined in $L_2(\gamma, E)$ as follows:

$$(u, v)_\gamma = \int_\gamma (u(Q), v(Q))_E \, ds \quad \left(ds^2 = \sum_{k=1}^{n} (dx^k)^2\right). \tag{2.1}$$

Suppose, further, that with each point Q in a domain of R_n there is associated a nondegenerate system of metrics $\mu_k(u, v; Q)$ $(k = 1, 2, \ldots, n)$ on E. We will also denote these metrics by $\mu_k(Q)$, omitting the arguments u and v.

By forming linear combinations $\sum_{k=1}^{n} \xi_k \mu_k(Q)$ of these metrics with real coefficients ξ_k, we obtain a linear space K(Q) of metrics on E. The metrics will be called the coordinate metrics.

An additional metric can now be introduced in the space $L_2(\gamma, E)$:

$$\mu(\gamma) = \mu(u, v; \gamma) = \int_\gamma \sum_{k=1}^{n} \mu_k(u(Q), v(Q), Q)\, dx^k = \int_\gamma (\vec{\mu} \cdot \vec{dr}) \quad (u, v \in L_2(\gamma, E), \tag{2.2}$$
$$\vec{\mu} = (\mu_1, \mu_2, \ldots, \mu_n), \ \vec{dr} = (dx^1, dx^2, \ldots, dx^n)).$$

(We require that $\vec{\mu}(Q) = (\mu_1(Q), \mu_2(Q), \ldots, \mu_n(Q))$ be a covariant vector in R_n so that the metrics $\mu(\gamma)$ will not depend on the choice of coordinate system in R_n.)

As is well known, a Hermitian form $\mu_k(Q)$ has the form

$$\mu_k(Q) = (\sigma_k(Q)\, u, v)_E \quad (u, v \in E), \tag{2.3}$$

where $\sigma_k(Q)$ $(k = 1, 2, \ldots, n)$ is a Hermitian operator in the Hilbert space E. It is required everywhere below that the mappings $\sigma_k(Q)$ be continuous functions of Q in the sense of the operator norm in E.

It is easily seen that

$$|\mu(u, v; \gamma)| \leqslant \sqrt{n} \sup_{(Q \in \gamma,\ k=1, 2, \ldots, n)} \| \sigma_k(Q) \| \, \| u \|_{L_2} \| v \|_{L_2}.$$

Consider the family of inductors

$$L(\gamma) = (H, \Phi(\gamma), L_2(\gamma, E), \mu(u, v; \gamma)), \tag{2.4}$$

where γ is an arbitrary piecewise-smooth arc curve belonging to a certain domain of R_n, the space H is a fixed Hilbert space and $\mu(u, v, \gamma)$ is defined by formula (2.2).

Definition. We will say that a family of inductors $L(\gamma)$ of form (2.4) *is regular at a point* Q_0 $(Q_0 \in R_n)$ if the following two conditions are satisfied.

1) For any (piecewise-smooth) arc γ emanating from Q_0, the mapping $\Phi(\gamma)$ takes any element $h \in H$ into a continuous E valued function $u(Q)$ $(Q \in \gamma)$ (an E-valued function is a function with values in E):

$$\lim_{Q' \to Q} \| u(Q') - u(Q) \|_E = 0 \quad (Q', Q \in \gamma). \tag{2.5}$$

2) There exists independently of γ a linear mapping $\dot{\Phi}(Q_0)$ of H into E such that

$$\| \Phi(\Delta\gamma) h |_Q - \Phi(Q_0) h \|_E \leqslant \varepsilon(\Delta\gamma) \| h \|, \quad (Q \in \Delta\gamma, \ h \in H), \tag{2.6}$$

where $\Delta\gamma$ is an arbitrary initial segment of an arc γ emanating from Q_0 and the symbol $|_Q$ means that the function is evaluated at Q (Figure 3). On each fixed arc γ we have $\lim \varepsilon(\Delta\gamma) = 0$ as the length Δs of $\Delta\gamma$ tends to zero.

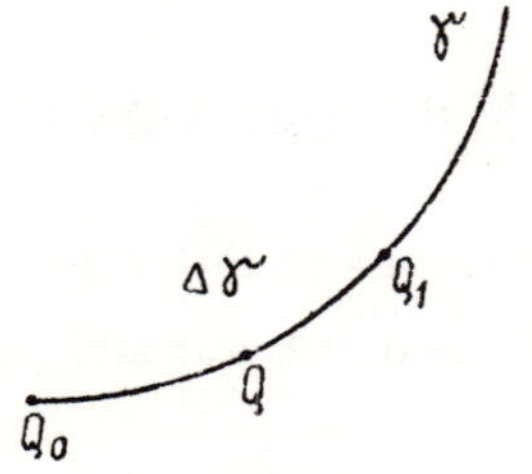

Fig. 3.

Definition. By the gradient at a point Q $(Q \in R_n)$ of a family of inductors $L(\gamma)$ that is regular at Q will be meant the n-metric inductor

$$\nabla_Q L = (H, \Phi(Q), E; \mu_1(Q), \ldots, \mu_n(Q)) \tag{2.7}$$

Analogously, by the *derivative* of a family of inductors $L(\gamma)$ that is regular at Q *in the direction of a vector* $\vec{a} = (a^1, a^2, \ldots, a^n)$ will be meant the monometric inductor

$$\frac{\partial L}{\partial \vec{a}} = \left(H, \dot{\Phi}(Q), E, \sum_{k=1}^{n} \mu_k(Q)\, a^k\right) \tag{2.8}$$

We note that the continuous mappings $\Phi(\gamma)$ $(H \xrightarrow{\Phi(\gamma)} L_2(\gamma, E))$ and $\dot{\Phi}(Q)$ $(H \xrightarrow{\dot{\Phi}(Q)} E)$ give rise to the adjoint continuous mappings $\Phi^+(\gamma)$ $(L_2(\gamma, E) \xrightarrow{\Phi^+(\gamma)} H)$ and $\dot{\Phi}_k^+(Q)$ $(E \xrightarrow{\dot{\Phi}_k^+(Q)} H)$ $(k = 1, 2, \ldots, n)$ satisfying the conditions

$$\mu(\Phi(\gamma)h, v; \gamma) = (h, \Phi^+(\gamma)v) \quad (v \in L_2(\gamma, E)), \tag{2.9}$$

$$\mu_k(\dot{\Phi}(Q)h, v; Q) = (h, \dot{\Phi}_k^+(Q)v) \quad (v \in E). \tag{2.10}$$

Lemma 2.1. Suppose $L(\gamma) = (H, \Phi(\gamma), L_2(\gamma, E), \mu(\gamma) = \int_\gamma (\vec{\mu} \cdot d\vec{r})$ is a family of inductors regular at a point Q_0, γ is a fixed arc emanating from Q_0 and $u(Q)$ is a continuous E-valued function on γ. Then on any initial segment $\Delta\gamma$ of γ,

$$\Phi^+(\Delta\gamma)\, u = \sum_{k=1}^{n} \dot{\Phi}_k^+(Q_0)\, u(Q_0)\, \Delta x^k + o(\Delta s), \tag{2.11}$$

where the Δx^k are the components of the vector $\overrightarrow{Q_0Q_1}$ joining the ends of the segment $\Delta\gamma$ and $\lim\limits_{\Delta s \to 0} \frac{o(\Delta s)}{\Delta s} = 0$. If h is an arbitrary element of H,

$$\Phi^+(\Delta\gamma)\, \Phi(\Delta\gamma)\, h = \sum_{k=1}^{n} \dot{\Phi}_k^+(Q_0)\, \dot{\Phi}(Q_0)\, h\, \Delta x^k + o_1(\Delta s, h), \tag{2.12}$$

where $\| o_1(\Delta s, h) \| \leqslant \delta(\Delta s)\, \| h \|\, \Delta s$ and $\lim\limits_{\Delta s \to 0} \delta(\Delta s) = 0$.

Proof. From equality (2.3), the continuity of the operators $\sigma_k(Q)$ and condition (2.6) we obtain the estimates

$$|\mu_k(u(Q),\ \Phi(\Delta\gamma)g|_Q;\ Q) - \mu_k(u(Q_0),\ \dot{\Phi}(Q_0)g;\ Q_0)| \leqslant |\mu_k(u(Q),\ \Phi(\Delta\gamma)g|_Q - \dot{\Phi}(Q_0)g;\ Q)| + |\mu_k(u(Q) - u(Q_0),\ \dot{\Phi}(Q_0)g;\ Q)| +$$
$$|\mu_k(u(Q_0),\ \dot{\Phi}(Q_0)g;\ Q) - \mu_k(u(Q_0),\ \dot{\Phi}(Q_0)g;\ Q_0)| \leqslant$$
$$C\|u(Q)\|\,\|\Phi(\Delta\gamma)g|_Q - \dot{\Phi}(Q_0)g\| + C\|u(Q) - u(Q_0)\| \times$$
$$\|\dot{\Phi}(Q_0)g\| + \|\sigma_k(Q) - \sigma_k(Q_0)\|\,\|u(Q_0)\|\,\|\dot{\Phi}(Q_0)g\| \leqslant$$
$$\varepsilon_1(\Delta s)\|g\| + \varepsilon_2(\Delta s)\|g\| + \varepsilon_3(\Delta s)\|g\| = \varepsilon(\Delta s)\|g\|,$$
$$(\lim_{\Delta s\to 0}\varepsilon(\Delta s) = 0,\ C = \sup_{(Q\in\gamma;\ k=1,\,2,\,\ldots,\,n)}\|\sigma_k(Q)\|). \tag{2.13}$$

By definition of the operator $\Phi^+(\Delta\gamma)$ we have

$$(\Phi^+(\Delta\gamma)u,\ g) = \mu(u,\ \Phi(\Delta\gamma)g;\ \Delta\gamma) = \int_{\Delta\gamma}\sum_{k=1}^{n}\mu_k(u(Q),\ \Phi(\Delta\gamma)h|_Q;\ Q)\,dx^k. \tag{2.14}$$

Integrating inequality (2.13) and using (2.14), we get

$$|(\Phi^+(\Delta\gamma)u,\ g) - \int_{\Delta\gamma}\sum_{k=1}^{n}\mu_k(u(Q_0),\ \dot{\Phi}(Q_0)g,\ Q_0)\,dx^k| \leqslant \sqrt{n}\,\varepsilon(\Delta s)\,\Delta s\,\|g\|,$$

which implies

$$|(\Phi^+(\Delta\gamma)u - \sum_{k=1}^{n}\dot{\Phi}_k^+(Q_0)\,u(Q_0)\,\Delta x^k,\ g)| \leqslant \sqrt{n}\,\varepsilon(\Delta s)\,\Delta s\,\|g\|. \tag{2.15}$$

Since g is an arbitrary element of H it follows from (2.15) that

$$\|\Phi^+(\Delta\gamma)u - \sum_{k=1}^{n}\dot{\Phi}_k^+(Q_0)\,u(Q)\,\Delta x^k\| \leqslant \sqrt{n}\,\varepsilon(\Delta s)\,\Delta s.$$

Relation (2.12) is proved analogously; it is only necessary to replace $u(Q)$ and $u(Q_0)$ in estimates (2.13)–(2.15) by $\Phi(\Delta\gamma)h$ and $\dot{\Phi}(Q_0)h$ respectively.

Theorem 2.1. Suppose $L(\gamma)$ is a family of inductors that is regular at a point Q_0, *γ is a fixed arc given by the equations $x^k = x^k(t)$ $(t_0 \leqslant t \leqslant t_1)$ and γ_λ is the* *initial segment of* γ *corresponding to the interval* $t_0 \leqslant t \leqslant \lambda$. *Then*

$$\operatorname{Ind} \frac{\partial L}{\partial \vec{a}_0} = \left. \frac{d \operatorname{Ind} L(\gamma_\lambda)}{d\lambda} \right|_{\lambda = t_0}, \tag{2.16}$$

where $\vec{a}_0 = (\dot{x}^1(t_0), \ldots, \dot{x}^n(t_0))$ *is the tangent vector to* γ *at* Q_0.

Proof. By virtue of (2.12) the induced metric has the form

$$\operatorname{Ind} L(\gamma_\lambda) = (\Phi^+(\gamma_\lambda)\Phi(\gamma_\lambda)h, h) = \sum_{k=1}^{n} (\dot{\Phi}_k^+(Q_0)\dot{\Phi}(Q_0)\Delta x^k h, h) + (0_1(\Delta s, h), h).$$

It follows that

$$\left. \frac{d \operatorname{Ind} L(\gamma_\lambda)}{d\lambda} \right|_{\lambda = t} = \lim_{\Delta\lambda \to 0} \frac{\operatorname{Ind} L(\gamma_\lambda)}{\Delta\lambda} = \sum_{k=1}^{n} (\dot{\Phi}_k^+(Q_0)\dot{\Phi}(Q_0)h, h)\dot{x}^k(t_0) =$$

$$\sum_{k=1}^{n} \mu_k (\dot{\Phi}(Q_0)h, \dot{\Phi}(Q_0)h)\dot{x}^k(t_0) = \operatorname{Ind} \frac{\partial L}{\partial \vec{a}_0}.$$

§2. DIFFERENTIATION OF FAMILIES OF METRIC COLLIGATIONS

1. Consider a family of metric colligations $M(\gamma) = (T(\gamma), L(\gamma))$, where $L(\gamma)$ is a family of inductors with a fixed internal space H and $T(\gamma)$ is a family of bounded linear operators in H.

We will say that the operators $T(\gamma)$ form a *regular family of internal operators at* $Q_0 \in R_n$ if there exists a system of bounded linear operators $A_1(Q_0), \ldots, A_n(Q_0)$ such that for any given piecewise-smooth arc γ emanating from Q_0 and any initial segment $\Delta\gamma$ of it, such that

$$T(\Delta\gamma) = I + \sum_{k=1}^{n} A_k(Q_0)\Delta x^k + o(\Delta s) \tag{2.17}$$

Definition. A family of metric colligations $M(\gamma) = (T(\gamma), L(\gamma))$ is said to be *regular* if both $T(\gamma)$ and $L(\gamma)$ are regular.

In this case we have the vector aggregate

$$\nabla_{Q_0} M(\gamma) = (A_1(Q_0), \ldots, A_n(Q_0), H, \dot{\Phi}(Q_0), E, \mu_1(Q), \ldots, \mu_n(Q)), \tag{2.18}$$

which is naturally called the *gradient at* Q_0 of the family of metric colligations $M(\gamma)$, while an aggregate of the form

$$\left.\frac{\partial M}{\partial a}\right|_{Q_0} = \left(\sum_{k=1}^{n} A_k(Q_0)\, a^k, \frac{\partial L}{\partial a}\right), \tag{2.19}$$

where $\vec{a} = (a^1, a^2, \ldots, a^n)$ is an arbitrary vector, is called the *derivative in the direction of* $\vec{a}$ of the family $M(\gamma)$.

Theorem 2.2. *If* $M(\gamma) = (T(\gamma), L(\gamma))$ *is a regular family of metric colligations at a point* Q_0 *then* $\left.\frac{\partial M}{\partial a}\right|_{Q_0}$ *is a local colligation and* $\nabla_{Q_0} M$ *is a vector local colligation.*

Proof. Consider a fixed arc γ emanating from Q_0 with tangent vector

$$\vec{\tau} = \vec{a} = (\dot{x}^1, \dot{x}^2, \ldots, \dot{x}^n)$$

at Q_0 (Fig. 4). Since $M(\gamma)$ is a metric colligation,

$$((T(\gamma(Q_0Q_t))\, T^*(\gamma(Q_0Q_t)) - I)\, h_1, h_2) = \operatorname{Ind} L(\gamma(Q_0Q_t)) \tag{2.20}$$

it follows by virtue of the regularity condition (2.17) and the rule (2.16) for differentiating induced metrics that

$$\frac{d}{dt}((T(\gamma(Q_0Q_t))\, T^*(\gamma(Q_0Q_t)) - I)\, h_1, \; h_2)\,|_{t=0} =$$
$$\left(\left\{\sum_{k=1}^{n} [A_k(Q_0) + A_k^*(Q_0)]\, a^k(Q_0)\right\} h_1, h_2\right) =$$
$$\frac{d}{dt} \operatorname{Ind} L(\gamma(Q_0Q_t))\Big|_{t=0} = \operatorname{Ind} \left.\frac{\partial L}{\partial \tau}\right|_{Q_0}. \tag{2.21}$$

Relations (2.21) imply that $\left.\frac{\partial M}{\partial a}\right|_{Q_0}$ is a local colligation.

Taking, in particular, the derivatives $\left.\frac{\partial M}{\partial x^k}\right|_{Q_0}$, we obtain the part of the theorem concerning $\nabla_{Q_0} M$.

2. A family of metric colligations $M(\gamma)$ will be called *multiplicative* if in a certain domain of R_n

$$M(\gamma_1 + \gamma_2) = M(\gamma_2)\, M(\gamma_1) \tag{2.22}$$

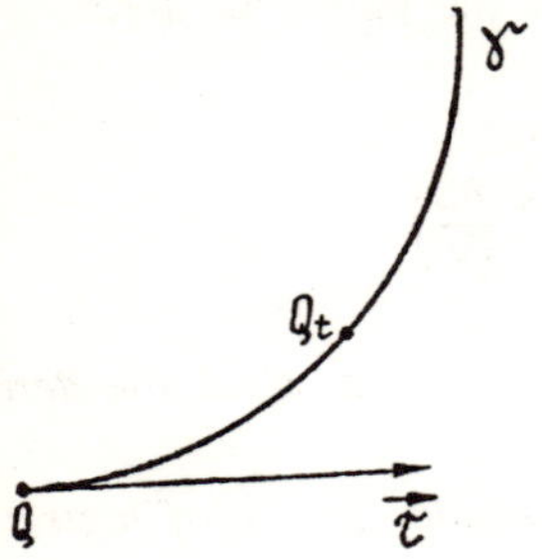

Fig. 4.

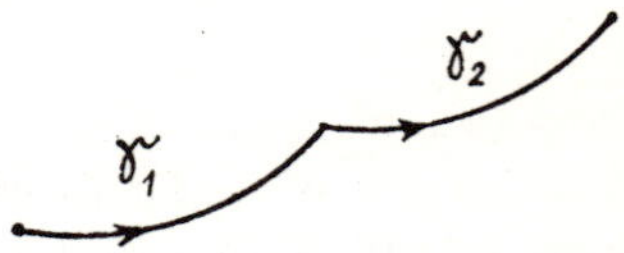

Fig. 5.

wherever γ_1 and γ_2 are piecewise-smooth arcs with the beginning of γ_2 coinciding with the end of γ_1, so that $\gamma_1 + \gamma_2$ is an arc composed of γ_1 and γ_2 (Fig. 5).

We note that $M(\gamma)$ has the form

$$M(\gamma) = (T(\gamma), H, \Phi(\gamma), L_2(\gamma, E), \mu(\gamma)), \tag{2.23}$$

whereas by virtue of (1.12)

$$M(\gamma_2) M(\gamma_1) = (T(\gamma_2) T(\gamma_1), H, \Phi(\gamma_1) T^*(\gamma_2) + \\ + \Phi(\gamma_2), L_2(\gamma_1 E) \dotplus L_2(\gamma_2, E), \mu(\gamma_1) + \mu(\gamma_2)). \tag{2.24}$$

Since $\gamma = \gamma_1 + \gamma_2$, there exists a natural isometry of $L_2(\gamma, E)$ onto $L_2(\gamma_1, E) \dotplus L_2(\gamma_2, E)$ that takes each E-valued function $u(Q)$ $(Q \in \gamma)$ into the pair of E-valued functions $\{u_1(Q_1), u_2(Q_2)\}$ $(Q_1 \in \gamma_1, Q_2 \in \gamma_2)$, where $u_1(Q_1) = u(Q_1)$ $(Q_1 \in \gamma_1)$ and $u_2(Q_2) = u(Q_2)$ $(Q_2 \in \gamma_2)$. The validity of the equality $M(\gamma_1 + \gamma_2) = M(\gamma_2) \dot{M}(\gamma_1)$ depends on this isometry between the spaces $L_2(\gamma, E)$ and $L_2(\gamma_1, E) \dotplus L_2(\gamma_2, E)$.

Theorem 2.3. *Let $M(\gamma) = (T(\gamma), H, \Phi(\gamma), L_2(\gamma, E), \mu(\gamma))$ be a regular multiplicative family of metric colligations, γ a fixed arc, and $u(Q)$ a continuous E-valued function defined on γ. Then the motion $g(t) = M(u(Q), \gamma(Q_0 Q_t))\, h$, which by virtue of* (1.10) *has the form*

$$g(t) = T(\gamma(Q_0 Q_t))\, h + \Phi^+(\gamma(Q_0 Q_t))\, u(Q) \tag{2.25}$$

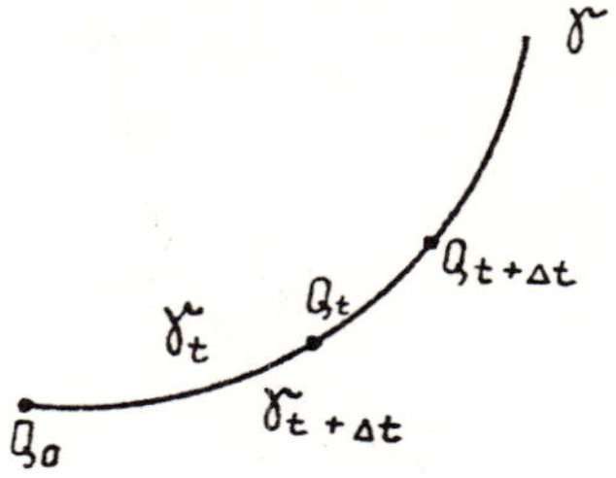

Fig. 6

is the solution of the initial-value problem

$$\frac{dg}{dt} = \Big(\sum_{k=1}^{n} A_k(Q_t)\,\dot{x}^k(t)\Big)\,g + \sum_{k=1}^{n} \Phi_k^+(Q_t)\,[u(Q_t)]\,\dot{x}^k(t),\ g(t_0) = h \tag{2.26}$$

We note that the right side of the differential equation of (2.26) is the motion with external influence $u(Q_t)$ corresponding to the colligation $\left.\frac{\partial M}{\partial \tau}\right|_{Q_t}$.

Proof. We have

$$g(t+\Delta t) = M(u;\ \gamma(Q_0Q_{t+\Delta t}))\,h = M(u,\gamma(Q_tQ_{t+\Delta t}))\,M(u,\gamma(Q_0Q_t))\,h = M(u,\gamma(Q_tQ_{t+\Delta t}))\,g(t) = T(\gamma(Q_tQ_{t+\Delta t}))\,g(t) + \Phi^+(\gamma(Q_tQ_{t+\Delta t}))\,u,$$

which implies

$$\frac{g(t+\Delta t)-g(t)}{\Delta t} = \frac{T(\gamma(Q_tQ_{t+\Delta t}))-I}{\Delta t}\,g(t) + \frac{\Phi^+(\gamma(Q_tQ_{t+\Delta t}))}{\Delta t}\,u. \tag{2.27}$$

By passing to the limit in (2.27) as $\Delta t \to 0$ and taking into account the regularity condition (2.17) and relation (2.11), we obtain the required initial-value problem (2.26).

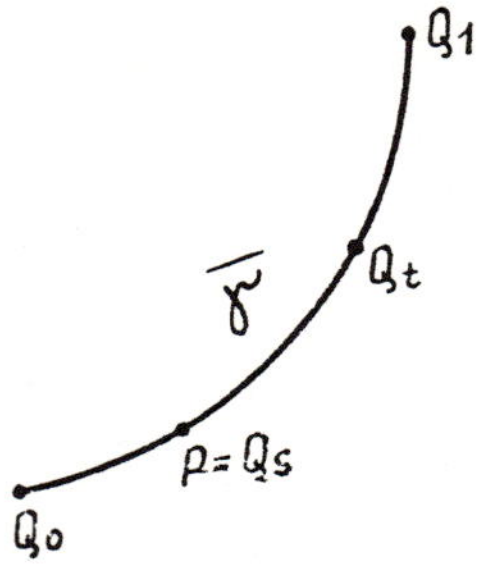

Fig. 7.

§ 3. INTEGRATION OF LOCAL COLLIGATIONS

In the preceding section it was shown that the operation of differentiation can be used to obtain a vector local colligation field $\nabla_Q M$ from a given regular multiplicative family of metric colligations $M(\gamma)$. Conversely, it turns out that a corresponding "integral" family of metric colligations can be constructed from a known vector local colligation field; namely, we have the following

Theorem 2.4. *Suppose a vector local colligation field* $X(Q) = (A_1(Q), \ldots, A_n(Q), H, \varphi, E, \mu_1(Q), \ldots, \mu_n(Q)$ *is given in a domain of* R_n. *If* $A_k(Q)$ $(k = 1, 2, \ldots, n)$ *and* $\varphi(Q))$ *are continuous functions of* Q *in the sense of the operator norm, there exists a regular multiplicative family of metric colligations* $M(\gamma) = (T(\gamma), H, \Phi(\gamma), L_2(\gamma, E), \mu(\gamma))$ *such that* $\nabla_Q M(\gamma) = X(Q)$.

Proof. From the field $X(Q)$ we obtain the following equations, which hold on any fixed arc γ (Fig. 7) emanating from a point Q_0:

$$\frac{dg}{dt} = \sum_{k=1}^{n} A_k(Q)\, \dot{x}^k g + \sum_{k=1}^{n} \varphi_k^+(Q)\,[u(Q)]\, \dot{x}^k \quad (t_0 \leqslant t \leqslant t_1),$$
$$g(Q_0) = h. \tag{2.28}$$

The solution of the initial-value problem (2.28) has the form

$$g(t) = W[\gamma(Q_0Q_t)]\,h + \int_{\gamma(Q_0Q_t)} W[\gamma(Q_sQ_t)] \sum_{k=1}^{n} \varphi_k^{+}(Q_s)[u(Q_s)]\,\dot{x}^k(s)\,ds \tag{2.29}$$

where the operator[1] $W[\gamma(Q_sQ_t)]$ in H is the solution of the initial-value problem

$$\frac{\partial W[\gamma(Q_sQ_t)]}{\partial t} = \sum_{k=1}^{n} A_k(Q_t)\,\dot{x}^k(t)\,W[\gamma(Q_sQ_t)], \quad W[\gamma(Q_sQ_s)] = I. \tag{2.30}$$

We note some properties of the operator $W[\gamma(Q_sQ_t)]$:

1) $W[\gamma(Q_1Q_3)] = W[\gamma(Q_2Q_3)]\,W[\gamma(Q_1Q_2)]$, (2.31)

where

$$\gamma(Q_1Q_3) = \gamma(Q_1Q_2) + \gamma(Q_2Q_3);$$

2) $W[\gamma(Q_1Q_2)]\,W[\gamma(Q_2Q_1)] = I$; (2.32)

3) $\dfrac{\partial W[\gamma(Q_sQ_t)]}{\partial s} = -W[\gamma(Q_sQ_t)] \sum_{k=1}^{n} A_k(Q_s)\,\dot{x}^k(s).$ (2.33)

Properties (2.31) and (2.32) are almost obvious, while property (2.33) is obtained by differentiating (2.32) and making use of (2.30).

Consider the following mapping of the Hilbert space H into the space $L_2(\gamma(Q_0Q_1), E)$:

$$\Phi(\gamma)\,h = \Phi(\gamma(Q_0Q_1))\,h \equiv \varphi(Q)\,W^*[\gamma(QQ_1)]\,h \quad (Q \in \gamma(Q_0Q_1)), \tag{2.34}$$

[1] The operator $W[\gamma(Q_sQ_t)]$ depends not only on the points Q_s and Q_t but also on the arc γ joining them. When it depends only on the points Q_t and Q_s, we can write $W(Q_t, Q_s)$, omitting the symbol γ.

where $\gamma(QQ_1)$ is a terminal segment of γ. By introducing the metric $\mu(\gamma) = \int\limits_{\gamma(Q_0Q_1)} \sum\limits_{k=1}^{n} \mu_k(Q)\, dx^k$ on the space $L_2(\gamma, E)$, we can obtain the following expression for the adjoint (relative to the metric $\mu(\gamma)$) operator $\Phi^+(\gamma)$:

$$\Phi^+[\gamma]\, u = \int\limits_{\gamma(Q_0Q_1)} W[\gamma(Q_tQ_1)] \sum_{k=1}^{n} \varphi_k^+(Q_t)\, \dot{x}^k(t)\, u(Q_t)\, dt. \tag{2.35}$$

In fact,

$$\mu(\Phi(\gamma)\, h, u; \gamma) = \int\limits_{\gamma(Q_0Q_1)} \sum_{k=1}^{n} \mu_k [\varphi(Q_t)\, W^*[\gamma(Q_tQ_1)]\, h, u]\, \dot{x}^k(t)\, dt =$$

$$\int\limits_{\gamma(Q_0Q_1)} \sum_{k=1}^{n} (h, W[\gamma(Q_tQ_1)]\, \varphi_k^+(Q_t)\, u)\, \dot{x}^k(t)\, dt.$$

Consider now the family of aggregates

$$M(\gamma) = (W[\gamma(Q_0Q_1)], H, \Phi(\gamma(Q_0Q_1)), L_2(\gamma, E), \mu(\gamma)). \tag{2.36}$$

We will show that this family of aggregates is the desired one. The regularity condition (2.17) for the internal operator $T(\gamma) = W[\gamma(Q_0Q_1)]$ follows from the initial-value problem (2.30), since

$$\lim_{\Delta t \to 0} \frac{W(\Delta\gamma)}{\Delta t} = \left.\frac{dW[\gamma(Q_{t_0}Q_t)]}{dt}\right|_{t=t_0} = \sum_{k=1}^{n} A_k(Q_{t_0})\, \dot{x}^k(t_0).$$

The inductor $(H, \Phi(\gamma(Q_0Q_1)), L_2(\gamma, E), \mu(\gamma))$ is also regular. In fact, the function $\Phi(\gamma(Q_0Q_1))\, h = \varphi(Q)\, W^*[\gamma(QQ_1)]\, h$ $(Q \in \gamma(Q_0Q_1))$ is a continuous E-valued function of Q. Moreover, the following estimates hold on any initial segment $\Delta\gamma$ $(t_0 \leqslant t \leqslant \lambda)$ of γ:

$$\|\Phi(\Delta\gamma)h|_{Q_t} - \varphi(Q_0)h\|_E = \|\varphi(Q_t)W^*(Q_tQ_\lambda)h - \varphi(Q_0)h\|_E \leqslant$$
$$\|\varphi(Q_t)(W^*(Q_tQ_\lambda) - I)h\|_E + \|\varphi(Q_t) - \varphi(Q_0)h\|_E \leqslant$$
$$C\|W^*(Q_tQ_\lambda) - I\| \cdot \|h\| + \delta(\lambda)\|h\|,$$
$$\lim_{\lambda \to t_0} \delta(\lambda) = 0,$$

where

$$C = \sup_{Q_t \in \gamma} \|\varphi(Q_t)\|.$$

It follows by virtue of the initial condition of (2.30) for $s = t_0$ that the mapping $\dot{\Phi}(Q_0)$ exists; in fact,

$$\dot{\Phi}(Q_0) = \varphi(Q_0). \tag{2.37}$$

The family $M(\gamma)$ is multiplicative. In fact, composing the product $M(\gamma_2)\,M(\gamma_1)$ and taking into account (2.31) and (2.34), we get

$$\begin{aligned} M(\gamma_2)\,M(\gamma_1) = (W[\gamma(Q_1Q_2)]\,W[\gamma(Q_0Q_1)],\ \Phi(\gamma(Q_1Q_2) \dotplus \\ \Phi(\gamma(Q_0Q_1))\,W^*[\gamma(Q_1Q_2)],\ L_2(\gamma_2, E) \dotplus L_2(\gamma_1, E), \mu(\gamma_1) + \\ \mu(\gamma_1) = (W[\gamma(Q_0Q_2)], \varphi(Q)\,W^*[\gamma(QQ_2)] \dotplus \\ \varphi(Q)\,W^*[\gamma(QQ_1)]\,W^*[\gamma(Q_1Q_2)], L_2(\gamma_2, E) \dotplus L_2(\gamma_1, E), \mu(\gamma)). \end{aligned} \tag{2.38}$$

Since

$$W^*[\gamma(QQ_1)]\,W^*[\gamma(Q_1Q_2)] = W^*[\gamma(QQ_2)],$$

by mapping the space $L_2(\gamma_1, E) \dotplus L_2(\gamma_2, E)$ onto the space $L_2(\gamma_1 + \gamma_2, E)$ in (2.38) by means of the natural isometry, we obtain the aggregate $(W[\gamma(Q_0Q_2)], \mathrm{H}, \varphi(Q)\,W^*[\gamma(QQ_2)], L_2(\gamma, E), \mu(\gamma))$, which coincides with $M(\gamma)$ $(\gamma = \gamma_1 + \gamma_2)$.

Let us show that $M(\gamma)$ is a metric colligation. From expressions (2.32) and (2.33) we get

$$\Phi^{+}(\gamma)\,\Phi(\gamma) = \int\limits_{\gamma(Q_0Q_1)} W[\gamma(Q_tQ_1)] \sum_{k=1}^{n} \varphi_k^{+}(Q_t)\,\varphi(Q_t)\,W^{*}[\gamma(Q_tQ_1)]\,\dot{x}^k(t)\,dt. \tag{2.39}$$

This result together with the local colligation conditions $\varphi_k^{+}(Q)\,\varphi(Q) = A_k(Q) + A_k^{*}(Q)$ implies

$$\Phi^{+}(\gamma)\,\Phi(\gamma) = \int\limits_{\gamma(Q_0Q_1)} W[\gamma(Q_tQ_1)] \sum_{k=1}^{n} A_k(Q_t)\,W^{*}[\gamma(Q_tQ_1)]\,\dot{x}^k(t)\,dt + \int\limits_{\gamma(Q_0Q_1)} W[\gamma(Q_tQ_1] \sum_{k=1}^{n} A_k^{*}(Q_t)\,W^{*}[\gamma(Q_tQ_1)]\,\dot{x}^k(t)\,dt. \tag{2.40}$$

Taking into account equation (2.33) and its adjoint equation, we get

$$\Phi^{+}(\gamma)\,\Phi(\gamma) = -\int\limits_{\gamma(Q_0Q_1)} \frac{d}{dt}\{W[\gamma(Q_tQ_1)]\}\,W^{*}[\gamma(Q_tQ_1)]\,dt - \int\limits_{\gamma(Q_0Q_1)} W[\gamma(Q_tQ_1)]\,\frac{d}{dt}\{W^{*}[\gamma(Q_tQ_1)]\}\,dt = -\int\limits_{\gamma(Q_0Q_1)} \frac{d}{dt}\{W[\gamma(Q_tQ_1)] \times W^{*}[\gamma(Q_tQ_1]\}\,dt = -I + W[\gamma(Q_0Q_1)]\,W^{*}[\gamma(Q_0Q_1)].$$

We have arrived at the metric colligation condition

$$\Phi^{+}(\gamma)\,\Phi(\gamma) = W[\gamma]\,W^{*}[\gamma] - I.$$

The theorem is completely proved.

It was not assumed in the preceding theorem that the system of metrics $\mu_1(Q), \ldots, \mu_n(Q)$ is nondegenerate. If it is nondegenerate, the following uniqueness theorem holds.

Theorem 2.5. If under the conditions of the preceding theorem the system of metrics $\mu_1(Q), \ldots \mu_n(Q)$ is nondegenerate in a domain of R_n, the integral regular multiplicative family of metric colligations $M(\gamma)$ satisfying the condition $\nabla_Q M(\gamma) = X(Q)$ is uniquely determined.

Proof. By considering the motions (1.10) corresponding to the colligations $M(\gamma)$ and using the equality $\nabla_Q M = X$, we obtain an initial-value problem of the form (2.26) in which $\dot{\Phi}_k^+(Q) = \varphi_k^+(Q)$. Since the solution of this problem is uniquely determined and has the form (2.29), the internal operator $T(\gamma) = W[\gamma(Q_0Q_1)]$ and the mapping

$$\Phi^+(\gamma(Q_0Q_1)) = \int_{\gamma(Q_0Q_1)} W[\gamma(Q_tQ_1)] \sum_{k=1}^{n} \varphi_k^+(Q_t)[\,\cdot\,]\dot{x}^k(t)\,dt$$

are uniquely determined. It remains for us to verify that the mapping $\Phi(\gamma)$ is also uniquely determined. For this purpose we need two observations:

1) If

$$\mu(u, v_0, \gamma) = \int_{\gamma(Q_0Q_1)} \sum_{k=1}^{n} \mu_k(u(Q_t), v_0, Q_t)\dot{x}^k(t)\,dt = 0, \tag{2.41}$$

where $u(Q)$ is any continuous E-valued function and $v_0(Q)$ is a fixed continuous function, then at each point $Q \in \gamma$

$$v_0(Q) \in \operatorname{Rad} \sum_{k=1}^{n} \mu_k \dot{x}^k \Big|_Q \tag{2.42}$$

(relation (2.42) is satisfied at corner points of γ by the limiting values from the left and from the right).

For suppose (2.42) is not satisfied. Then there exist at least one arc γ, one point $\bar{Q} \in \gamma$ and one element $u_0 \in E$ such that

$$\sum_{k=1}^{n} \mu_k(u_0, v_0(\bar{Q}), \bar{Q})\dot{x}^k(\bar{Q}) \neq 0.$$

If one introduces an E-valued function of the form $u(Q) = u_0 c(Q)$, where $c(Q)$ is a continuous scalar function of Q, and makes use of the continuity of the metrics $\mu_k(Q)$, one can easily show that the middle part of (2.41) can be made different from zero; in fact, it suffices to take

$$c(Q) = \sum_{k=1}^{n} \mu_k (u_0, v_0(Q), Q) \dot{x}^k (Q).$$

2) If $\gamma = \gamma_1 + \gamma_2$, the values on γ_2 of the continuous function $\Phi(\gamma) h = u(h, Q, \gamma)$ do not depend on γ_1 (Fig. 8).

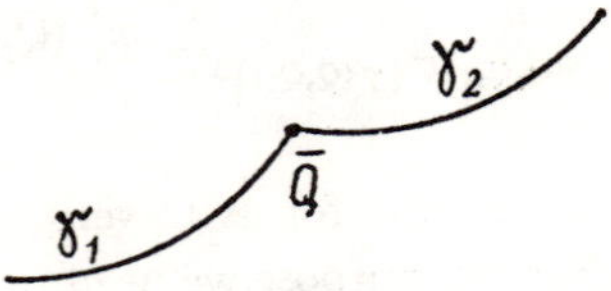

Fig. 8

In fact, since

$$M(\gamma_1 + \gamma_2) = M(\gamma_2) M(\gamma_1) = (W[\gamma_2] W[\gamma_1], H, \Phi(\gamma_1) W^*[\gamma_2] \dot{+} \\ \dot{+} \Phi(\gamma_2), L_2(\gamma_1, E) \dot{+} L_2(\gamma_2, E), \mu(\gamma_1) + \mu(\gamma_2)),$$

it follows that

$$\Phi(\gamma) h = \begin{cases} \Phi(\gamma_2) h, & Q \in \gamma_2, \\ \Phi(\gamma_1) W^*[\gamma_2] h, & Q \in \gamma_1 \end{cases} \tag{2.43}$$

and hence $\Phi(\gamma) h$ $(Q \in \gamma_2)$ does not depend on γ_1.

Suppose now there exist two mappings $\Phi'(\gamma)$ and $\Phi''(\gamma)$ such that $\Phi'^{+}(\gamma) = \Phi''^{+}(\gamma)$ and let $v_0(Q) = \Phi'(\gamma) h - \Phi''(\gamma) h$. Then

$$\mu(u, v_0; \gamma) = ((\Phi'^{+}(\gamma) - \Phi''^{+}(\gamma))(u, h) = 0$$

and hence

$$v_0(Q) \in \mathrm{Rad} \sum_{k=1}^{n} \mu_k \dot{x}^k |_Q . \tag{2.44}$$

We fix a point $\bar{Q}$ on γ and accordingly divide γ into two segments: $\gamma(QQ_1) = \gamma(Q_0\bar{Q}) + \gamma(\bar{Q}Q_1)$. The values $\dot{x}^k(\bar{Q}-0)$ can be chosen arbitrarily by varying the segment $\gamma(Q_0\bar{Q})$. But then relation (2.44) implies

$$v_0(Q) \in \bigcap_{k=1}^{n} \operatorname{Rad}\mu_k(\bar{Q}) = 0.$$

Thus $\Phi'(\gamma)h = \Phi''(\gamma)h$. Q.E.D.

Remark. The results obtained in this section can be used to introduce the notion of a product of regular multiplicative families

$$M^{(k)}(\gamma) = (T^{(k)}(\gamma),\ H^{(k)},\ \Phi^{(k)}(\gamma),\ L_2(\gamma,\ E),\ \mu(\gamma)) \quad (k = 1,2)$$

of metric colligations with a common external space.

To this end we find $\nabla_Q M^{(k)}(\gamma)$ and then form the E product $X(Q) = \nabla_Q M^{(1)}(\gamma) \vee \nabla_Q M^{(2)}(\gamma)$ in accordance with §4 of Chapter I. Since an E-product of local colligations is a local colligation, $X(Q)$ is a field of vector local colligations. We form the corresponding integral family $M(\gamma)$ of metric colligations and call it the E-product, written $M(\gamma) = M^{(1)}(\gamma) \vee M^{(2)}(\gamma)$, of the families of metric colligations $M^{(1)}(\gamma)$ and $M^{(2)}(\gamma)$. It is obvious that an E-product of regular multiplicative families of metric colligations is again a regular multiplicative family of metric colligations.

NOTE

The results of this chapter are due to M.S. Livshits.

CHAPTER III

THE ASSOCIATED OPEN SYSTEMS

§1. THE EQUATIONS OF OPEN SYSTEMS

1. Let γ be a piecewise smooth arc in R_n and let H and E be two Hilbert spaces.

Definition. By an *open system* F_γ *on* γ is meant a pair of continuous linear mappings

$$H \dotplus L_2(\gamma, E) \xrightarrow{R_\gamma} L_2(\gamma, H),$$
$$H \dotplus L_2(\gamma, E) \xrightarrow{S_\gamma} H \dotplus L_2(\gamma, E). \tag{3.1}$$

The elements $\{h_0, u(Q)\}$ $(h_0 \in H;\ u(Q) \in L_2(\gamma, E);\ Q \in \gamma)$ of $H \dotplus L_2(\gamma, E)$ are called the *inputs* of the system F_γ, while their images $h(Q) = R_\gamma\{h_0, u(Q)\}$ and $\{h, v(Q)\} = S_\gamma\{h_0, u(Q)\}$ are respectively called the *internal states and outputs* of the system F_γ.

The operator S_γ is called the *transition* (or *transmission) operator* of F_γ. The components h_0 and h_1 are called the *internal input* and *internal output*, while the components $u(Q)$ and $v(Q)$ are called the *external input* and *external output* respectively.

Let $M(\gamma) = (T(\gamma), H, \Phi(\gamma), L_2(\gamma, E), \mu(\gamma))$ be a regular multiplicative family of metric colligations and let $\nabla_Q M = (A_1(Q), \ldots, A_n(Q), H, \varphi(Q), E, \mu_1(Q), \ldots, \mu_n(Q))$, where $\varphi(Q) = \dot{\Phi}(Q)$, be its gradient field.

Definition. An open system F_γ defined on a piecewise smooth arc lying in some domain of R_n is said to be *associated with the regular multiplicative family of metric colligations* $M(\gamma)$ (or with the local colligation field $\nabla_Q M$) if the mappings R_γ and S_γ satisfy the conditions

$$h(Q) = R_\gamma\{h_0, u(Q)\} = T(\gamma(Q_0Q))\,h_0 + \Phi^+(\gamma(Q_0Q))\,u, \tag{3.2}$$

$$v(Q) = u(Q) + \Phi(Q)\,h(Q) \quad (Q \in \gamma), \tag{3.3}$$

$$h_1 = h(Q_1). \tag{3.4}$$

It clearly follows from relation (3.2) and the conditions for the regularity of $M(\gamma)$ that

$$h_0 = h(Q_0) \tag{3.5}$$

It will be shown in §3 of this chapter that a given linear physical system of general form with appropriate connecting channels can always be regarded as a system associated with some regular multiplicative family of metric colligations.

We note that in the case of a fixed external influence u ($u \in L_2(\gamma, E)$) and a fixed point Q the mapping (3.2) defines a motion in H of the form (1.10) with colligation $M(\gamma(Q_0Q))$. Since $M(\gamma)$ is a multiplicative family, there corresponds to the sum $\gamma(Q_0Q_1) + \gamma(Q_1Q_2)$ a successive application of two motions (from Q_0 to Q_1, and then from Q_1 to Q_2) with external influences $u_1(Q)$ $(Q \in \gamma(Q_0Q_1))$ and $u_2(Q)$ $(Q \in \gamma(Q_1Q_2))$.

2. By virtue of Theorem 2.3 the internal state (3.2) of an associated system is the solution of the initial problem

$$\frac{dh}{dt} = \sum_{k=1}^{n} A_k(Q_t)\dot{x}^k(t)\,h + \sum_{k=1}^{n} \varphi_k^+(Q_t)\,[u(Q_t)]\,\dot{x}^k(t), \tag{3.6}$$
$$h(0) = h(Q_0).$$

For open systems we have the following

Theorem 3.1. *If F_γ is an associated open system and u_1, u_2 are two inputs, while v_1, v_2 and h_1, h_2 are the corresponding outputs and internal states, then*

$$\int_\gamma \sum_{k=1}^{n} [\mu_k (v_1 (Q), v_2 (Q)) - \mu_k (u_1 (Q), u_2 (Q))]\, dx^k = (h_1 (Q_1), h_2 (Q_1)) - (h_1 (Q_0), h_2 (Q_0)). \tag{3.7}$$

Proof. Condition (3.3) implies

$$\mu (v_1, v_2, \gamma) = \mu (u_1, u_2, \gamma) + \mu (u_1, \varphi h_2, \gamma) + \mu (\varphi h_1, h_2, \gamma) + \mu (\varphi h_1, \varphi h_2, \gamma), \tag{3.8}$$

where

$$\mu (u, v, \gamma) = \int_\gamma \sum_{k=1}^{n} \mu_k (u (Q), v (Q), Q)\, dx^k.$$

In addition, it follows from (3.6) that the following relation holds on the arc γ:

$$\frac{d (h_1, h_2)}{dt} = \left(\sum_{k=1}^{n} A_k \dot{x}^k h_1, h_2\right) + \left(\sum_{k=1}^{n} \overset{+}{\varphi}_k [u_1] \dot{x}^k, h_2\right) + \left(h_1, \sum_{k=1}^{n} A_k \dot{x}^k h_2\right) + \left(h_1, \sum_{k=1}^{n} \overset{+}{\varphi}_k [u_2] \dot{x}^k\right). \tag{3.9}$$

The local colligation conditions $A_k + A_k^* = \overset{+}{\varphi}_k \varphi$ can be used to rewrite this relation in the form

$$\frac{d (h_1, h_2)}{dt} = \sum_{k=1}^{n} (\overset{+}{\varphi}_k \varphi h_1, h_2)\, \dot{x}^k + \sum_{k=1}^{n} \mu_k (u_1, \varphi h_2, Q)\, \dot{x}^k + \sum_{k=1}^{n} \mu_k (\varphi h_1, u_2, Q)\, \dot{x}^k. \tag{3.10}$$

Integrating (3.10) along γ and comparing the result with (3.8), we obtain the required relation (3.7).

Corollary (metric conservation principle). *In the special case* $u_1 = u_2$, $v_1 = v_2$, $h_1 = h_2$ *relation* (3.7) *takes the form*

$$\int_\gamma \sum_{k=1}^{n} [\mu_k (v(Q), v(Q), Q) - \mu_k (u(Q), u(Q), Q)] \, dx^k = \| h(Q_1) \|^2 - \| h(Q_0) \|^2. \tag{3.11}$$

2. Let $X(Q) = (A_1(Q), \ldots, A_n(Q), H, \varphi(Q), E, \mu_1(Q), \ldots, \mu_n(Q))$ be a vector local colligation field. Clearly, the aggregate

$$X'(Q) = (-A_1^*(Q), \ldots, -A_n^*(Q), H, \varphi(Q), E, -\mu_1(Q), \ldots, -\mu_n(Q)).$$

is also a vector local colligation field. The fields $X(Q)$ and $X'(Q)$ will be called *dual* relative to each other.

The corresponding integral families of metric colligations $M(\gamma)$ and $M'(\gamma)$ will also be called *dual*.

In accordance with equality (2.36) $M(\gamma)$ has the form

$$M(\gamma) = (W[\gamma], H, \Phi(\gamma), L_2(\gamma, E), \mu(\gamma)). \tag{3.12}$$

Here $W[\gamma]$ is the solution of problem (2.30) while the mapping $\Phi(\gamma)$ is defined by formula (2.34):

$$\Phi(\gamma) h = \varphi(Q) W^* [\gamma(QQ_1)] h \quad (Q \in \gamma), \tag{3.13}$$

where $\gamma(QQ_1)$ is a terminal segment of γ.

From the definition of the family $M'(\gamma) = (W'[\gamma], H, \Phi'(\gamma), L_2(\gamma, E), \mu'(\gamma) = -\mu(\gamma))$ and property (2.32) when applied to the adjoint operator it can be seen that

$$W'[\gamma(Q_0Q_1)] = W^*[\gamma(Q_1Q_0)]. \tag{3.14}$$

It therefore follows from (3.13) that

$$\Phi'(\gamma)\,h = \varphi(Q)\,W[\gamma(Q_1Q)]\,h. \tag{3.15}$$

An open system $F'(\gamma')$ associated with the family of colligations $M'(\gamma')$ is said to be *dual* relative to an open system $F(\gamma)$ associated with the family $M(\gamma)$.

Since $\varphi' = \varphi$ and $\varphi_k'^{+} = -\varphi_k^{+}$, the equations of the system $F'(\gamma')$ have the form

$$\frac{dh'}{dt'} = -\sum_{k=1}^{n} A_k^*(Q_{t'}')\,\dot{x}^k(Q_{t'}')\,h'(Q_{t'}') - \sum_{k=1}^{n} \varphi_k^{+}(Q_{t'}')\,u'(Q_{t'}')\,x^k(Q_{t'}'),$$
$$h'|_{t'=t_0'} = h'(Q_{t_0'}') \qquad (t_0' < t' < t_1'), \tag{3.16}$$

$$h_1' = h'(Q_{t'}'),$$
$$v'(Q') = u'(Q') + \varphi(Q')\,h'(Q') \quad (Q' \in \gamma'). \tag{3.17}$$

The following theorem establishes the relationship between the inputs, outputs and internal states of certain pairs of dual open systems $F(\gamma)$ and $F'(\gamma')$.

Theorem 3.2. *Suppose that the arc* $\gamma'(Q_0'Q_1') = \gamma(Q_1Q_0)$ $(Q_0' = Q_1,\ Q_1' = Q_0)$ *differs from the arc* $\gamma(Q_0Q_1)$ *only in the direction of motion. If the input of the system* $F'(\gamma')$ *is given by the equalities*

$$h'(Q_0') = -h(Q_1),\quad u'(Q) = v(Q) \quad (Q \in \gamma'), \tag{3.18}$$

where $\{h(Q_1),\ v(Q)\}$ *is the output of the system* $F(\gamma)$, *then the output and internal state of the system* $F'(\gamma')$ *are given by the equalities*

$$h'(Q_1') = -h(Q_0),\quad v'(Q) = u(Q) \quad (Q \in \gamma'), \tag{3.19}$$

$$h'(Q) = -h(Q), \tag{3.20}$$

where $\{h(Q_0),\ u(Q)\}$ *and* $h(Q)$ *are the corresponding input and internal state respectively of the system* $F(\gamma)$.

Proof. From the differential equation of (3.6) and the local colligation conditions $A_k + A_k^* = \varphi_k^+ \varphi$ we have

$$\frac{dh}{dt} = -\sum_{k=1}^{n} A_k^* (Q_t) \dot{x}^k (Q_t) h(Q_t) + \sum_{k=1}^{n} \varphi_k^+ (Q_t) [u (Q_t) + \varphi (Q_t) h (Q_t)] \dot{x}^k (Q_t).$$

This result together with equality (3.3) implies

$$\frac{dh}{dt} = -\sum_{k=1}^{n} A_k^* (Q_t) \dot{x}^k (Q_t) h (Q_t) + \sum_{k=1}^{n} \varphi_k^+ (Q_t) v (Q_t) \dot{x}^k (Q_t), \quad (3.21)$$

$$h|_{t=t_0} = h (Q_0). \tag{3.21}$$

In order to obtain relation (3.20) we make the substitutions $t' = t_0 - t + t_1$, $h' (Q'_{t'}) = -h (Q_t)$ $(Q'_{t'} = Q_t)$ in (3.21). Then by virtue of the equalities $\varphi' (Q) = \varphi (Q)$, $\varphi_k'^+ (Q) = -\varphi_k^+ (Q)$ problem (3.21) goes over into problem (3.16) for determining the internal state of the system $F' (\gamma')$.

Equality (3.3) implies that the input $u (Q)$ has the form

$$u (Q) = v (Q) - \varphi (Q) h (Q) = u' (Q) + \varphi' (Q) h' (Q) = v' (Q),$$

in agreement with (3.19).

Theorem 3.3. *The transition operator S_γ of an associated system $F (\gamma)$ effects an automorphism of the space $H \dotplus L_2 (\gamma, E)$ with*

$$S_\gamma^{-1} = \hat{I} S'_{\gamma'} \hat{I} \tag{3.22}$$

where $S'_{\gamma'}$ is the transition operator of a dual system, γ' is the arc differing from γ only in the direction of motion, and $\hat{I}$ is the operator such that

$$\hat{I} \{h, u\} = \{-h, u\}. \tag{3.23}$$

The operator S_γ is a unitary operator in $H \dotplus L_2 (\gamma, E)$ relative to the metric *defined by the equality*

$$\mu(\{h, u\}, \{h, u\}; \gamma) = \|h\|^2 - \int_\gamma \sum_{k=1}^{n} \mu_k (u(Q), u(Q), Q)\, dx^k. \quad (3.24)$$

The proof follows directly from Theorems 3.1 and 3.2.

§2. COUPLING AND DECOMPOSITION OF OPEN SYSTEMS

Suppose we are given, on an arc $\gamma(Q_0Q_1)$, a set of open systems $F_\gamma^{(k)}$ ($k = 1, 2, \ldots, m$) with internal spaces $H^{(1)}, H^{(2)}, \ldots, H^{(m)}$ and a common external space $L_2(\gamma, E)$. Let $R_\gamma^{(k)}$ and $S_\gamma^{(k)}$ ($k = 1, 2, \ldots, m$) be the corresponding input to internal state and input to output mappings of these systems. We extend (with preservation of linearity) the definition of the linear mappings $R_\gamma^{(k)}$ and $S_\gamma^{(k)}$ onto the whole space $H \dotplus L_2(\gamma, E)$, where $H = H^{(1)} \oplus H^{(2)} \oplus \ldots \oplus H^{(m)}$, by putting

$$R_\gamma^{(k)} \{h_j, 0\} = \{0, 0\} \quad k \neq j, \quad (3.25)$$

$$S_\gamma^{(k)} \{h_j, 0\} = \{h_j, 0\} \quad k \neq j. \quad (3.26)$$

Definition. An open system F_γ is called a *coupling* of the systems $F_\gamma^{(k)}$ ($k = 1, 2, \ldots, m$) and is written in the form $F_\gamma = F_\gamma^{(1)} \vee F_\gamma^{(2)} \vee \ldots \vee F_\gamma^{(m)}$ if its internal space has the form $H = H^{(1)} \oplus H^{(2)} \oplus \ldots \oplus H^{(m)}$, its external space coincides with $L_2(\gamma, E)$ and its mappings R_γ and S_γ are defined by the equalities

$$R_\gamma = R_\gamma^{(1)} + R_\gamma^{(2)} S_\gamma^{(1)} + \ldots + R_\gamma^{(m)} S_\gamma^{(m-1)} \ldots S_\gamma^{(1)}, \quad (3.27)$$

$$S_\gamma = S_\gamma^{(m)} S_\gamma^{(m-1)} \ldots S_\gamma^{(2)} S_\gamma^{(1)}. \quad (3.28)$$

The definitions (3.27) and (3.28) of the mappings R_γ and S_γ imply that under a coupling of open systems the external output of the preceding system becomes the external input of the system under consideration (Fig. 9), while the corresponding internal states, internal inputs and internal outputs of different systems are added together as orthogonal elements.

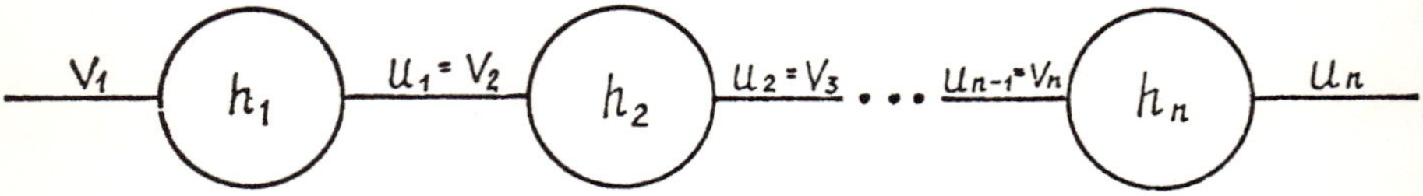

Fig. 9.

We note that the operation of coupling of open systems is associative:

$$F_{\Upsilon}^{(1)} \vee (F_{\Upsilon}^{(2)} \vee F_{\Upsilon}^{(3)}) = (F_{\Upsilon}^{(1)} \vee F_{\Upsilon}^{(2)}) \vee F_{\Upsilon}^{(3)}.$$

Theorem 3.4. *Suppose we are given a set of vector local colligation fields* $X^{(k)}(Q) = (\vec{A}^{(k)}(Q),\ H^{(k)},\ \varphi^{(k)}(Q),\ E,\ \vec{\mu}(Q))\ (k = 1, 2, \ldots, m)$, *where the vector notation* $\vec{A} = (A_1, A_2, \ldots, A_n)$ *and* $\vec{\mu} = (\mu_1, \mu_2, \ldots, \mu_n)$ *denotes the corresponding n-tuples of operators and metrics, and consider a corresponding set of open systems* $F_{\Upsilon}^{(k)}$ $(k = 1, 2, \ldots, m)$ *that are respectively associated with them. Then the coupling* $F_{\Upsilon} = F_{\Upsilon}^{(1)} \vee F_{\Upsilon}^{(2)} \vee \ldots \vee F_{\Upsilon}^{(m)}$ *is a system associated with the E-product (Chapter I, §4)* $X(Q) = X^{(1)}(Q) \vee X^{(2)}(Q) \vee \ldots \vee X^{(m)}(Q)$ *of the local colligation fields* $X^{(k)}(Q)$.

Proof. It suffices to consider the case $m = 2$. The initial-value problem (3.6) for a system $\hat{F}_{\Upsilon}$ associated with the field $X(Q) = X^{(1)}(Q) \vee X^{(2)}(Q)$ has the form

$$\begin{gathered} \frac{dh}{dt} = \left(\vec{A} \cdot \frac{d\vec{r}}{dt}\right) h + \left(\vec{\varphi}^{+} \cdot \frac{d\vec{r}}{dt}\right) u, \\ h \,|_{\, t=t_0} = h(Q_0), \end{gathered} \tag{3.29}$$

where $\vec{A} \cdot d\vec{r}/dt$ and $\vec{\varphi}^{+} \cdot d\vec{r}/dt$ denote the scalar products $\sum_{k=1}^{n} A_k(Q_t) \frac{dx^k}{dt}$ and

$\sum_{k=1}^{n} \varphi_k^+ (Q_t) \frac{dx^k}{dt}$ of the vectors $\vec{A} = (A_1, A_2, \ldots, A_n)$ and $\vec{\varphi}^+ = (\varphi_1^+, \varphi_2^+, \ldots, \varphi_n^+)$ times the vector $\frac{\vec{dr}}{dt} = \left(\frac{dx^1}{dt}, \ldots, \frac{dx^n}{dt}\right)$ respectively. Since $X(Q) = X^{(1)}(Q) \vee X^{(2)}(Q)$, it follows by virtue of conditions (1.14), (1.15) that

$$\vec{A} = \vec{A}^{(1)} P^{(1)} + \vec{A}^{(2)} P^{(2)} + \vec{\varphi}^{(2)+} \varphi^{(1)} P^{(1)}, \tag{3.30}$$

$$\varphi = \varphi^{(1)} P^{(1)} + \varphi^{(2)} P^{(2)}, \tag{3.31}$$

$$\vec{\varphi}^+ = \vec{\varphi}^{(1)+} + \vec{\varphi}^{(2)+}. \tag{3.32}$$

Projecting both sides of the equations (3.29) onto the subspaces $H^{(1)}$ and $H^{(2)}$ and using equalities (3.30) and (3.32), we get

$$\begin{gathered} \frac{dh^{(1)}}{dt} = \left(\vec{A}^{(1)} \cdot \frac{\vec{dr}}{dt}\right) h^{(1)} + \left(\vec{\varphi}^{(1)+} \cdot \frac{\vec{dr}}{dt}\right) u; \\ h^{(1)}/_{t=t_0} = P^{(1)} h(Q_0); \end{gathered} \tag{3.33}$$

$$\begin{gathered} \frac{dh^{(2)}}{dt} = \left(\vec{A}^{(2)} \cdot \frac{\vec{dr}}{dt}\right) h^{(2)} + \left(\vec{\varphi}^{(2)+} \cdot \frac{\vec{dr}}{dt}\right) \left(u + \vec{\varphi}^{(1)} h^{(1)}\right); \\ h^{(2)}/_{t=t_0} = P^{(2)} h(Q_0). \end{gathered} \tag{3.34}$$

From these equations we see that the internal state h is the sum of the internal states $h^{(1)} = R^{(1)} \{P^{(1)} h(Q_0), u\}$ and $h^{(2)} = R^{(2)} \{P^{(2)} h(Q_0), u + \varphi^{(1)} h^{(1)}\}$. But by virtue of equality (3.3) the output of the first system $F^{(1)}$ has the form $\{P^{(1)} h(Q_1), u + \varphi^{(1)} h^{(1)}\}$, which implies the equality

$$h = h^{(1)} + h^{(2)} = (R^{(1)} + R^{(2)} S^{(1)}) \{h(Q_0), u\} \tag{3.35}$$

Relation (3.35) means that the mapping $\hat{R}$ for the system $\hat{F}_\gamma$ coincides with the mapping $R = R^{(1)} + R^{(2)} S^{(1)}$ for the coupling $F_\gamma = F_\gamma^{(1)} \vee F_\gamma^{(2)}$.

Further, by virtue of relations (3.3), (3.31) and (3.35), for the external output of the system $\hat{F}_\gamma$ we have

$$v = u + \varphi h = u + (\varphi^{(1)}P^{(1)} + \varphi^{(2)}P^{(2)})(h^{(1)} + h^{(2)}) = \\ u + \varphi^{(1)}h^{(1)} + \varphi^{(2)}h^{(2)} = v^{(1)} + \varphi^{(2)}R^{(2)}\{h^{(2)}(Q_0),\ v^{(1)}\}. \tag{3.36}$$

The right side of (3.36) is an expression of form (3.3) for the external output of the system $F_\gamma^{(2)}$, the external input of which is the external output $v^{(1)}$ of the system $F_\gamma^{(1)}$.

It follows that $\hat{S} = S_2S_1$. The theorem is proved.

Corollary. *If $H = H_0 \supset H_1 \supset \ldots \supset H_{m-1} \supset H_m = 0$ is a chain of invariant subspaces of the internal operators $\vec{A}(Q)$ for all Q of a vector local colligation field $X(Q)$, then an open system F_γ associated with $X(Q)$ decomposes into the coupling $F_\gamma = F_\gamma^{(1)} \vee F_\gamma^{(2)} \vee \ldots \vee F_\gamma^{(m)}$ of open systems $F_\gamma^{(j)}$ ($j = 1, 2, \ldots, m$) associated with the local colligations fields $X_j^\perp(Q) = P_j^\perp X(Q)$, where $P_j^\perp$ is the orthogonal projection onto the subspace $H_j^\perp = H_{j-1} \ominus H_j$* ($j = 1, 2, \ldots, m$) (see Chapter I, §4).

We can obtain a corresponding decomposition of a regular multiplicative family $M(\gamma)$ of metric colligations if the internal operators $T(\gamma)$ have a chain $H = H_0 \supset H_1 \supset \ldots \supset H_{m-1} \supset H_m = 0$ of invariant subspaces not depending on γ by going over to the gradient field $\nabla_Q M$ and making use of Theorems 3.4 and 2.4.

§3. GENERAL LINEAR SYSTEMS AND THEIR CONNECTION WITH ASSOCIATED OPEN SYSTEMS

1. In chapter I the notion of a vector local colligation

$$X = (A_1, \ldots, A_n, H, \varphi, E, \mu_1, \ldots, \mu_n)$$

was introduced with the use of conditions on the real parts of the internal operators A_k. But in many cases it is more convenient to deal with the imaginary parts of the operators A_k by using local colligation conditions of the form

$$\frac{1}{i}(A_k - \overset{+}{A}_k) = 2\mathrm{Im}A_k = \overset{+}{\varphi}_k\varphi. \tag{3.37}$$

In order to pass from a vector local colligation satisfying conditions of the form (1.9) to one satisfying conditions (3.37), it suffices to replace each operator A_k by iA_k, the mapping φ by $-i\varphi$ and each metric μ_k by $-\mu_k$. Each mapping $\overset{+}{\varphi}_k$ is then replaced by $-i\overset{+}{\varphi}_k$ while the basic equations (3.6), (3.3) of an associated open system become

$$i\frac{dh}{dt}+\left(\sum_{k=1}^{n}A_k\dot{x}^k(t)\right)h=\sum_{k=1}^{n}\overset{+}{\varphi}_k[u]\,\dot{x}^k(t), \tag{3.38}$$
$$h(0)=h_0,$$

$$v=u-i\varphi h. \tag{3.39}$$

Also, the transmission operator S_γ becomes a unitary operator relative to the metric

$$\mu(\{h,u\},\{h,u\})=\|h\|^2+\int_\gamma\sum_{k=1}^{n}\mu_k(u(Q),\ u(Q),Q)\,dx^k \tag{3.40}$$

which is obtained from the metric (3.24) by replacing each μ_k by $-\mu_k$.

2. In accordance with the general theory of linear physical systems [2], [42] the motion of a system as a function of time is described by equations of the form

$$i\frac{dh}{dt}+Ah=K[\xi_1(t)], \tag{3.41}$$
$$h(0)=h_0,$$

$$\xi_2(t)=Lh, \tag{3.42}$$

where ξ_1, ξ_2, h are respectively the input, output and internal state of the system, the values $\xi_1(t)$, $\xi_2(t)$, $h(t)$ of which belong to Hilbert spaces E_1, E_2, H respectively, A is a bounded linear operator in H, K is a linear operator mapping E_1 into H and L is a linear operator mapping H into E_2.

In order to construct a local colligation with which system (3.41), (3.42) can be associated, we first include the operator A in a local colligation $(A, H, \psi_3, E_3, \mu_3)$ so that

$$\frac{1}{i}(A - A^*) = \psi_3^+ \psi_3 \cdot$$

Let $\psi_1 = K^*$ and $\psi_2 = iL$, where K^* denotes the adjoint of K relative to the Hilbert metrics of H and E_1.

We next introduce the space $E = E_1 \dot{+} E_1 \dot{+} E_2 \dot{+} E_2 \dot{+} E_3$ with metric

$$\mu(u', u'') = (\xi_1', \eta_1'') + (\eta_1', \xi_1'') + (\xi_2', \eta_2'') + (\eta_2', \xi_2'') + \mu_3(\xi_3', \xi_3''),$$

where the elements u have the form $u = \{\xi_1, \eta_1, \xi_2, \eta_2, \xi_3\}(\xi_1, \eta_1 \in E_1; \xi_2, \eta_2 \in E_2; \xi_3 \in E_3)$ and the parentheses denote the scalar products in E_1 and E_2.

We define a mapping φ of H into E by putting

$$\varphi h = (0, \psi_1 h, 0, \psi_2 h, \psi_3 h) \cdot \tag{3.43}$$

Then the adjoint φ^+ of φ relative to the metric μ has the form

$$\varphi^+ u = \psi_1^* \xi_1 + \psi_2^* \xi_2 + \psi_3^+ \xi_3 \tag{3.44}$$

since

$$\mu(u, \varphi h) = (\xi_1, \psi_1 h) + (\xi_2, \psi_2 h) + \mu_3(\xi_3\ \psi_3 h) =$$
$$(\psi_1^* \xi_1, h) + (\psi_2^* \xi_2, h) + (\psi_3^+ \xi_3, h).$$

It follows from (3.43) and (3.44) that $\varphi^+ \ \varphi = \psi_3^+ \psi_3 = \frac{1}{i}(A - A^*)$. Thus the aggregate (A, H, φ, E, μ) is a local operator colligation. The equations of an open system F associated with this colligation have the form

$$i\frac{dh}{dt} + Ah = \varphi^+ u, \tag{3.45}$$

$$h(0) = h_0,$$
$$v = u - i\varphi h. \qquad (3.46)$$

Setting $u = (\xi_1, \eta_1, \xi_2, \eta_2, \xi_3)$, $v = (\tilde{\xi}_1, \tilde{\eta}_1, \tilde{\xi}_2, \tilde{\eta}_2, \tilde{\xi}_3)$, we get

$$i\frac{dh}{dt} + Ah = \psi_1^*\xi_1 + \psi_2^*\xi_2 + \psi_3^*\xi_3; \qquad (3.47)$$
$$h(0) = h_0,$$

$$\tilde{\xi}_1 = \xi_1, \quad \tilde{\eta}_1 = \eta_1 - i\psi_1 h;$$
$$\tilde{\xi}_2 = \xi_2, \quad \tilde{\eta}_2 = \eta_2 - i\psi_2 h; \qquad (3.48)$$
$$\tilde{\xi}_3 = \xi_3 - i\psi_3 h.$$

If now the input of F is an element of the form $u = (\xi_1(t), 0, 0, 0, 0)$, equations (3.47), (3.48) reduce to the equations

$$i\frac{dh}{dt} + Ah = K\xi_1(t);$$
$$h(0) = h_0; \qquad (3.49)$$
$$\tilde{\xi}_1 = \xi_1(t), \quad \tilde{\eta}_1 = -iK^*h;$$

$$\tilde{\xi}_2 = 0, \quad \tilde{\eta}_2 = Lh; \qquad (3.50)$$
$$\tilde{\xi}_3 = -i\psi_3 h.$$

From (3.49) and (3.50) we see that under the above special form of the input the internal state of an associated system F coincides with the internal state of the original linear system while the component $\tilde{\eta}_2$ of the output of F coincides with the output of the original system.

Thus, *for any linear system of form* (3.41), (3.42) *there exists an external space E relative to which this system can be regarded as associated with some local colligation* (or with some regular multiplicative family of metric colligations).

§4. RIEMANNIAN GEOMETRY AND TENSOR COLLIGATIONS[1]

Let D be a domain of a Euclidean space R_n and let $ds^2 = g_{ij}dx^i dx^j$ be the square of the element of arc length in an arbitrary curvilinear coordinate system. Suppose that at each point $Q(x^1, \ldots, x^n)$ of the domain D there is given a connection object $\Gamma_{ij}^k(Q), Q \in D$, where the functions $\Gamma_{ij}^k(Q) = \Gamma_{ij}^k(x_1, \ldots, x_n)$ are differentiable a sufficient number of times. Then the infinitesimal parallel displacement of a vector $h = (h^1, \ldots, h^n)$ is defined by[2]

$$dh^k = -\Gamma_{ij}^k h^j dx^i. \tag{3.51}$$

By a vector h we mean a contravariant vector field $h^i(Q)$ whose value at a point Q is interpreted as a vector $h(Q)$ in the affine tangent space H_Q. We do not require that the parallel displacement (of a vector) defined by (3.51) coincide with parallel displacement in the Euclidean space R_n, although this special case is not excluded.

Thus the domain D, when considered as an n-dimensional manifold, is provided with two structures, viz. that of the Euclidean space R_n and that of an affine connection space L_n with connection object Γ_{ij}^k [35].

A scalar product can be introduced in the space H_Q by putting

$$(h_1, h_2) = g_{ij}(Q) h_1^i h_2^j, \tag{3.52}$$

where

$$h_\alpha = h_\alpha^i e_i \quad (\alpha = 1, 2)$$

in which the e_i belong to a local frame of the affine space H_Q. Clearly, $(e_i, e_j) = g_{ij}(Q)$.

[1] The material of this section is not needed for an understanding of the sequel.

[2] In which the repeating-index summation convention is used.

Since R_n is a Euclidean space, there exists a rectangular coordinate system in which $ds^2 = (dx^i)^2$. In this special coordinate system the scalar product in H_Q has the form

$$(h_1, h_2)_Q = h_1^i h_2^i \quad (e_i, e_j) = \delta_{ij} \tag{3.53}$$

and it can be assumed that the space $H_Q = H$ does not depend on the point Q.

We introduce operators A_i $(i = 1, 2, \ldots, n)$ in H by putting

$$(A_i h)^k = -\Gamma_{ij}^k h^j. \tag{3.54}$$

Then the equations for parallel displacement along a curve $\gamma(Q_0Q_1)$ $(x^k = x^k(t),\ (t_0 \leqslant t \leqslant t_1))$ can be written in the form

$$\frac{dh}{dt} = \left(A_i \frac{dx_i}{dt}\right) h,$$
$$h(t_0) = h(Q_0). \tag{3.55}$$

The operators $T(\gamma)$ determining the parallel displacement $h(Q_1) = T(\gamma)\, h(Q_0)$ along curves γ in L_n clearly form a regular multiplicative family in the sense of Chapter II, §2, with

$$T(\gamma(Q_0Q_1)) = I + A_i \Delta x^i + o(\Delta s)$$

where the A_i are defined by (3.54).

Since parallel displacements in R_n and L_n do not coincide, the operators $T(\gamma)$ are not unitary in the metric (3.53) of H: an observer in R_n reckons that the metric is not conserved under a parallel displacement in L_n. He is justified in explaining this as due to an interaction with the external world and he can construct an associated open system, identifying the internal state of this system with parallel displacement in L_n under the assumption that the external input $u(Q)$ $(Q \in \gamma)$ of the system is equal to zero. The external output of the system will then be different from zero. To obtain this system one must include the operators $A_1(Q), \ldots, A_n(Q)$ in a vector local colligation field $X(Q) = (A_1(Q), \ldots, A_n(Q), H, \varphi(Q), E, \mu_1(Q), \ldots, \mu_n(Q))$ with a nondegenerate set of metrics $\mu_k(Q) : \bigcap_{k=1}^{n} \operatorname{Rad} \mu_k(Q) = 0$.

Consider the symmetric operators $\sigma_i = A_i + A_i^*$ defined in H by the matrices $\sigma_i = \| \sigma_{ij}^k \|$, where

$$\sigma_{ij}^k = -(\Gamma_{ij}^k + \Gamma_{ik}^j). \tag{3.56}$$

Let $H_0(Q) = H \ominus \bigcap_{i=1}^n \operatorname{Ker}\sigma_i(Q)$ and suppose that $r_0 = \dim H_0(Q)$ does not depend on Q in D. According to Theorem 1.1, the set of metrics $(\sigma_i(Q)\, h,\ h)$ in H can be included in a polymetric inductor $(H,\ \varphi'(Q),\ H_0(Q),\ \mu_1'(Q), \ldots, \mu_n'(Q))$ so that

$$(\sigma_i h_1,\ h_2) = \mu_i'(\varphi'(Q)\, h_1,\ \varphi'(Q)\, h_2),$$

where $\varphi'(Q) = P_0(Q)$ is the orthogonal projection of H onto $H_0(Q)$. Since the $\sigma_i(Q)$ are continuous functions of Q and $r_0 = \dim H_0(Q)$ is constant, $P_0(Q)$ is also a continuous function of Q. We take a fixed space E and a one-to-one linear mapping $K(Q)$ of $H_0(Q)$ onto E that depends continuously on Q $(Q \in D)$.

Setting $\mu_i(u,\ v) = \mu_i'(K^{-1}u,\ K^{-1}v)$, we obtain a nondegenerate inductor $(H,\ \varphi = K(Q)\,\varphi'(Q),\ E,\ \mu_1(Q), \ldots, \mu_n(Q))$ with fixed E such that $\sigma_i = \operatorname{Ind} \mu_i$.

We have thus shown that *if* $\dim \bigcap_{i=1}^n \operatorname{Ker}\sigma_i(Q)$ *is a constant quantity, the operators* $A_i(Q)$ *constructed from the connection object* $\Gamma_{ij}^k(Q)$ *by formulas* (3.54) *can be included in a nondegenerate local colligation field*

$$X(Q) = (A_1(Q), \ldots, A_n(Q),\ H,\ \varphi(Q),\ E,\ \mu_1(Q), \ldots, \mu_n(Q))$$

In terms of the basis elements the local colligation condition has the form

$$(\sigma_i e_j,\ e_k) = -(\Gamma_{ij}^k + \Gamma_{ik}^j) = \mu_i(\varphi e_j,\ \varphi e_k). \tag{3.57}$$

With the use of the field $X(Q)$ we can now write the equations of the associated open system on an arc γ (Fig. 10):

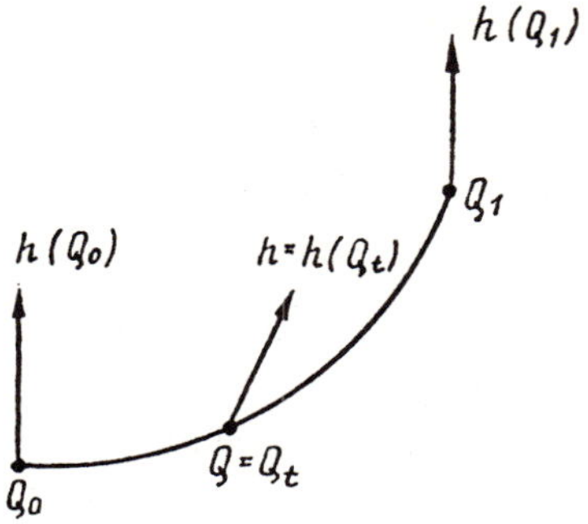

Fig. 10.

$$\frac{dh^k}{dt} + \Gamma^k_{ij} h^j \frac{dx^i}{dt} = [\varphi_i^+(u)]^k \frac{dx^i}{dt} \quad (t_0 \leqslant t \leqslant t_1), \tag{3.58}$$

$$h^k|_{t=t_0} = h^k(Q_0),$$

$$v(Q) = u(Q) + \varphi(Q)\,h(Q) \quad (u(Q),\ v(Q) \in E). \tag{3.59}$$

The mapping $\varphi_i^+(u)$ of E into H is determined from the condition

$$\mu_i(\varphi h,\ u) = (h,\ \varphi_i^+(u)) = h^k[\varphi_i^+(u)]^k \quad (u \in E), \tag{3.60}$$

where

$$\varphi_i^+(u) = [\varphi_i^+(u)]^k e_k \quad (e_k \in H).$$

But the local colligation condition (3.57) and equations (3.58), (3.59) are only valid in a special coordinate system in which $ds^2 = (dx^i)^2$. Let us find the form of the local colligation condition in an arbitrary curvilinear coordinate system. Under a coordinate transformation $x'^i = f^i(x^1, \ldots, x^n)$ the components h^i and the local frame elements are transformed according to the rules

$$h'^i = \frac{dx'^i}{\partial x^j} h^j, \tag{3.61}$$

$$e'_i = \frac{\partial x_j}{\partial x'^i} e_j . \tag{3.62}$$

Since $\mu_i \, dx^i$ is invariant, the metrics $\mu_i (u, v)$ $(u, v \in E)$ form a covariant vector and the mapping $\varphi h = u$ does not depend on the choice of coordinate system.

Suppose now x^i is an arbitrary curvilinear coordinate system in the Euclidean space R_n and e_i is a local frame at a point Q. Then any vector in the tangent space H_Q has the form $h = h^i e_i$.

The scalar product in H_Q is given by equality (3.52).

Theorem 3.5. *In any curvilinear coordinate system the local colligation condition* (3.57) *has the form*

$$\Delta_i g_{jk} = \mu_i (\varphi e_j, \ \varphi e_k), \tag{3.63}$$

where

$$\Delta_l g_{jk} = \frac{\partial g_{jk}}{\partial x^l} - \Gamma^s_{lj} g_{sk} - \Gamma^s_{lk} g_{js} \tag{3.64}$$

is the covariant derivative of the tensor g_{jk} *in the space* L_n *with connection object* Γ^k_{ij}.

Proof. It is well known that the covariant derivative $\Delta_i g_{jk}$ is a covariant tensor of degree three. In a rectangular coordinate system $g_{ij} = \delta_{ij}$ and hence the expression $\Delta_i g_{jk} = -(\Gamma^k_{ij} + \Gamma^j_{ik})$ coincides with the left side of condition (3.63). If we show that in any coordinate system the quantities $\mu_i (\varphi e_j, \ \varphi e_k)$ form a tensor that is covariant in all of its indices, equality (3.63) will have been proved.

Setting

$$\theta_{ijk} = \mu_i (\varphi e_j, \ \varphi e_k), \tag{3.65}$$

we get

$$\theta_{ijk} h^i_1 h^j_2 \, dx^k = \mu_k (\varphi h_1, \ \varphi h_2) \, dx^k, \tag{3.66}$$

where $h_\alpha = h^i_\alpha e_i \quad (\alpha = 1, \ 2)$. Since φh_1 and φh_2 are elements of E that do not depend on the choice of coordinate system, the right side of (3.66) is invariant and, inasmuch as h_1^l, h_2^l, dx^k are arbitrary contravariant vectors, the quantities θ_{ilk} form a covariant tensor of degree three.

We note that the open system equations (3.58), (3.59) retain their form in any coordinate system if the mappings $\overset{+}{\varphi_i}(u)$ of E into H_Q are determined from the conditions

$$\mu_i(\varphi h,\ u) = (h,\ \varphi_i^+(u))_Q = g_{jk}(Q)\, h^j\, [\varphi_i^+(u)]^k. \tag{3.67}$$

In fact, the left side of (3.58) is the covariant derivative of h^k in the direction of the tangent to γ and is therefore a contravariant vector. From (3.66) it follows that the expression

$$\mu_i(\varphi h,\ u)\, dx^i = g_{jk} h^j\, [\varphi_i^+(u)]^k\, dx^i$$

is invariant, and since h^j and dx^k are arbitrary contravariant vectors, the quantities $[\varphi_i^+(u)]^k$ form a tensor that is covariant in i and contravariant in k. This implies that the expression $[\varphi_i^+(u)]^k \frac{dx^i}{dt}$ is a contravariant vector. Inasmuch as both sides of (3.58) coincide in one coordinate system, they also coincide in an arbitrary coordinate system.

We have deduced the colligation condition (3.57) and equations (3.58) of an associated open system under the assumption that the metric tensor $g_{ij}(Q)$ $(Q \in D)$ defines a Euclidean geometry in D. Suppose now that the tensor $g_{ij}(Q)$ defines an arbitrary Riemannian geometry V_n in D. Then it is possible to postulate condition (3.57) and equations (3.58) for a space L_n/V_n provided with two geometries: the Riemannian geometry V_n with metric tensor g_{ij} and the affine connection geometry L_n with connection object Γ_{ij}^k. Equations (3.58) of an associated open system pertain to an observer who makes a change in the geometry V_n and watches a parallel displacement in L_n. In particular, the geometry L_n itself can be Riemannian with metric tensor $\overline{g_{ij}}$. The connection object Γ_{ij}^k in this case coincides with the Christoffel symbol $\overline{\Gamma}_{ij}^k$ for the tensor $\overline{g_{ij}}$, while the metric $\overline{g_{ij}} h^i h^j$ is conserved under a parallel displacement in L_n.

Definition. An aggregate

$$X(Q) = (\Gamma^k_{ij}(Q),\ H_Q,\ \varphi(Q),\ E,\ \mu_1(Q),\ \ldots,\ \mu_n(Q))$$

in which the operators $A_l(Q)$ are defined by means of a connection object $\Gamma^k_{lj}(Q)$ is called a *tensor colligation of class* L_n/V_n if conditions (3.63) are satisfied:

$$\Delta_l g_{jk} = \mu_l(\varphi e_j,\ \varphi e_k).$$

Theorem 3.6. *For an open system defined by equations* (3.58) *and associated with a tensor colligation of class* L_n/V_n *the conservation principle*

$$\mu_\Gamma(v,\ v) - \mu_\Gamma(u,\ u) = (g_{\alpha\beta}h^\alpha h^\beta)_{Q_1} - (g_{\alpha\beta}h^\alpha h^\beta)_{Q_0} \tag{3.68}$$

holds for the metric

$$\mu_\Gamma(u,\ u) = \int_\Gamma \mu_k(u,\ u)\, dx^k \tag{3.69}$$

in $L_2(\Gamma,\ E)$. *Here* Γ *is an arc with ends* Q_0 *and* Q_1 *while* $u(Q)$, $v(Q)$ *and* $h(Q)$ $(Q \in \Gamma)$ *are respectively the external input, external output and internal state of the open system* F_Γ.

Proof. Consider the differential

$$d(g_{jk}h^jh^k) = \frac{\partial}{\partial x^l}(g_{jk}h^jh^k)\, dx^l = \frac{\partial g_{jk}}{\partial x^l} h^jh^k\, dx^l + g_{jk}\frac{\partial h^j}{\partial x^l} h^k\, dx^l + g_{jk}h^j\frac{\partial h^k}{\partial x^l}\, dx^l.$$

With the use of the open system equations (3.58) we get

$$d(g_{ij}h^jh^k) = \frac{\partial g_{jk}}{\partial x^l} h^jh^k\, dx^l - g_{jk}\Gamma^j_{il}h^i\, dx^l h^k -$$
$$g_{jk}h^j\Gamma^k_{il}h^i\, dx^l + g_{jk}[\varphi^+_l(u)]^j\, dx^l h^k + g_{jk}[\varphi^+_l(u)]^k\, dx^l h^j.$$

It follows from the tensor colligation condition (3.63) and relation (3.67) for the adjoint mapping $\varphi^+_l(u)$ that

$$d\,(g_{ij}h^jh^k) = \frac{\partial g_{jk}}{\partial x^l}\,h^jh^k\,dx^l - \Gamma^s_{lj}g_{sk}h^jh^k\,dx^l -$$
$$\Gamma^s_{lk}g_{js}h^jh^k\,dx^l + 2\mu_l\,(\varphi h,\ u)\,dx^l = \Delta_l g_{jk}h^jh^k\,dx^l +$$
$$2\mu_l\,(\varphi h,\ u)\,dx^l = [\mu_l\,(\varphi h,\ \varphi h) + 2\mu_l\,(\varphi h,\ u)]\,dx^l. \tag{3.70}$$

But from the equality $v = u + \varphi h$ we have

$$\mu_l\,(v,\ v) = \mu_l\,(u,\ u) + 2\mu_l\,(\varphi h,\ u) + \mu_l\,(\varphi h,\ \varphi h), \tag{3.71}$$

and this result together with (3.70) implies

$$d\,(g_{ij}h^jh^k) = \mu_l\,(v,\ v)\,dx^l - \mu_l\,(u,\ u)\,dx^l. \tag{3.72}$$

The required equality (3.68) follows upon integrating (3.72) along Γ.

Example. Consider a one-dimensional manifold with two Riemannian metrics

$$ds^2 = g_{11}\,(x)\,dx^2,\quad d\bar{s}^2 = \bar{g}_{11}\,(x)\,dx^2. \tag{3.73}$$

The ratio $\frac{\bar{g}_{11}\,(x)}{g_{11}\,(x)}$ is clearly invariant relative to a coordinate transformation $x' = f\,(x)$. In the special case under consideration the Christoffel symbols

$$\bar{\Gamma}^l_{jk} = \bar{g}^{\alpha l}\bar{\Gamma}_{\alpha,\,jk},$$
$$\bar{\Gamma}_{\alpha,\,jk} = \frac{1}{2}\left(\frac{\partial \bar{g}_{\alpha j}}{\partial x^k} + \frac{\partial \bar{g}_{\alpha k}}{\partial x^j} - \frac{\partial \bar{g}_{jk}}{\partial x^\alpha}\right)$$

reduce to the single function

$$\bar{\Gamma}^1_{11} = \frac{1}{2}\,\bar{g}_{11}^{-1}\,\frac{d\bar{g}_{11}}{dx} = \frac{1}{2}\,\frac{d}{dx}\ln \bar{g}_{11} \tag{3.74}$$

Since $\dim H = 1$, any vector of H has the form $h = h'e_1$. The tensor θ_{ljk} reduces to the expression $\theta_{111} = \mu_1\,(\varphi e_1,\ \varphi e_1)$. We define a mapping $\varphi\,(H \overset{\varphi}{\to} E)$ by putting

$$\varphi e_1 = \varphi_1 a, \tag{3.75}$$

where $\boldsymbol{a}$ is the unit vector of E ($\dim E = 1$) and φ_1 is a numerical function of $\boldsymbol{x}$. From the condition $\overline{\Delta}_1 g_{11} = \theta_{111}$ we have

$$\frac{dg_{11}}{dx} - \overline{g}_{11}^{-1}\frac{d\overline{g}_{11}}{dx} g_{11} = \mu_1(\varphi e_1, \varphi e_1) = \varphi_1^2 \mu_1(a, a),$$

which implies

$$\mu_1(a, a) = \varphi_1^{-2} g_{11} \frac{d}{dx} \ln \frac{g_{11}}{\overline{g}_{11}}. \tag{3.76}$$

The mapping $\varphi^+(u) = \xi'(u)\, e_1$ is determined from the condition

$$\mu_1(\varphi h, u) = g_{11} h' \xi'(u).$$

Since $\boldsymbol{u} = (u, a)\, \boldsymbol{a}$,

$$\varphi^+(u) = \varphi^+(a)(u, a) = (u, a)\, \xi'(a)\, e_1, \tag{3.77}$$

$$\begin{aligned}\mu_1(\varphi h, u) = \mu_1(h' \varphi e_1, u) = h'(u, a)\, \varphi_1 \mu_1(a, a) = \\ g_{11} h'(u, a)\, \xi'(a)\end{aligned} \tag{3.78}$$

and hence

$$\xi'(a) = g_{11}^{-1} \varphi_1 \mu_1(a, a) = \varphi_1^{-1} \frac{d}{dx} \ln \frac{g_{11}}{\overline{g}_{11}}. \tag{3.79}$$

Equations (3.58) of the associated open system in the present case will be

$$\frac{dh'}{dx} + \frac{1}{2}\left(\frac{d}{dx} \ln \overline{g}_{11}\right) h' = \varphi_1^{-1} \left(\frac{d}{dx} \ln \frac{g_{11}}{\overline{g}_{11}}\right) u, \tag{3.80}$$

$$v = u + \varphi_1 h', \tag{3.81}$$

where for the sake of brevity we have denoted (u, a) and (v, a) by u and v respectively. Equations (3.80 and (3.81) can be reduced by means of simple transformations to the form

$$\frac{d\psi}{ds} = \frac{d}{ds}(\ln q)\,\frac{u}{\rho},$$
$$v = u + \rho\psi, \tag{3.82}$$

where we have introduced the notation

$$\psi = h'\sqrt{\bar{g}_{11}},\ \rho = \frac{\varphi_1}{\sqrt{\bar{g}_{11}}},\ q = \frac{\bar{g}_{11}}{g_{11}}.$$

We note that the quantities ψ, ρ and q are invariant relative to a transformation $x' = f(x)$. By replacing $\frac{u}{\rho}$ and $\frac{v}{\rho}$ by u and v respectively, we can take $\rho = 1$.

The external metric μ_Γ has the form

$$\mu_\Gamma(u, u) = \int_\Gamma \mu_1(a, a)\, u^2\, dx = \int_\Gamma u^2\, dq. \tag{3.83}$$

The open system equations (3.82) have a simple geometric meaning. Suppose that $q = \left(\frac{ds}{d\bar{s}}\right)^2 < 1$ and set

$$q = \left(1 + \left(\frac{dy}{ds}\right)^2\right)^{-1}. \tag{3.84}$$

We find the function $y = y(s)$ from (3.84) and construct its graph on the (s, y) plane (Fig. 11).

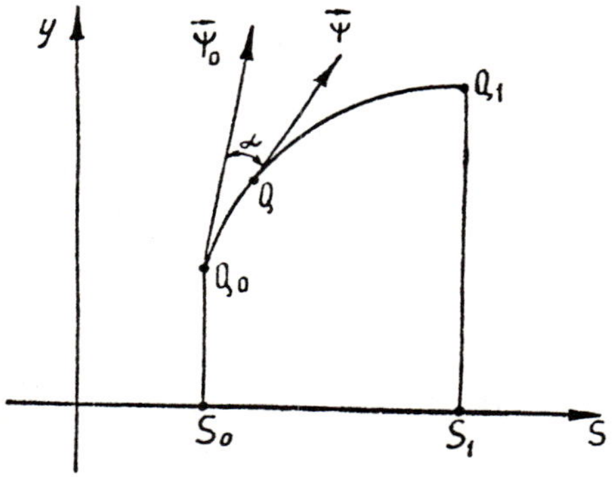

Fig. 11.

It is easily seen that

$$\frac{d}{ds}\ln q = K, \tag{3.85}$$

where $K = -\frac{d^2y}{ds^2}\left(1+\left(\frac{dy}{ds}\right)^2\right)^{-3/2}$ is the curvature of the curve $y = y(s)$. If in the open system equations

$$\begin{aligned} \frac{d\psi}{ds} &= Ku, \\ v &= u + \psi \end{aligned} \tag{3.86}$$

we set the input $u = 0$, we obtain in accordance with the general theory of equation (3.86) a parallel displacement in the geometry of $d\bar{s}^2 = \bar{g}_{11}\, dx^2$. Since $d\bar{s}^2 = \left(1+\left(\frac{dy}{ds}\right)^2\right) ds^2$, this displacement coincides with the motion of the tangent vector $\vec{\psi}$ along the curve $y = y(s)$.

Since $K = \frac{d\alpha}{ds}$, where α is the turning angle of the tangent, the quantity $u = \frac{d\psi}{d\alpha}$ measures the rate of change of the length of the vector $\vec{\psi}$ relative to the turning angle of the tangent. The input to internal state and input to output mappings for the open system under consideration have the forms

$$\begin{aligned} \psi(\alpha) &= \psi(0) + \int_0^\alpha u(\lambda)\, d\lambda \quad (0 \leqslant \alpha \leqslant \alpha_1), \\ v(\alpha) &= u(\alpha) + \int_0^\alpha u(\lambda)\, d\lambda + \psi(0). \end{aligned} \tag{3.87}$$

NOTES

The notion of a linear system as a pair of mappings R and S (input to internal state and input to output respectively) was introduced in the works of Zadeh and Desoer [42] and others [30]. The problem of constructing a general theory of linear systems in Hilbert and Banach spaces has also been investigated by

Balakrishnan [2[1,2]], who obtained the equations for a linear system by imposing certain restrictions on the mappings R and S.

But the operators appearing in these equations were not connected by the operator colligation condition, which left it impossible to carry out a subsequent analysis (the decomposition of a system into a "chain" of elementary systems, the coupling of systems, etc.).

The results of §3, which are due to Livshits and Yantsevich, establish a connection between arbitrary linear systems and the open systems associated with operator colligations and permit one to study linear systems with the methods of the theory of non-self-adjoint operators.

The results of §4 are due to M.S. Livshits. If a Riemannian space V_n and an inductor field $(H_q, \varphi(q), E, \mu_1(q), \ldots, \mu_n(q))$, ($q \in V_n$ and H_q is the tangent space at q) are given, there exists a unique symmetric affine connection without torsion satisfying the tensor colligation condition (3.63). For, by permuting the cyclic indices in (3.63) and (3.64), one can readily obtain the equalities

$$\Gamma_{k,ij} = \Gamma_{ij}{}^{p} g_{pk} = \frac{1}{2}\left(\frac{\partial g_{ik}}{\partial x^j} + \frac{\partial g_{jk}}{\partial x^i} - \frac{\partial g_{ij}}{\partial x^k}\right) + \frac{1}{2}(\mu_{k,ij} - \mu_{j,ik} - \mu_{i,jk}), \quad (1)$$

where $\mu_{k,ij} = \mu_k(\varphi e_i, \varphi e_j)$.

The simplest example of the space L_n/V_n is obtained by putting $V_n = R_n$ (the Euclidean space of dimension n) and $\dim E = 1$. One next defines the metrics μ_i and the mapping φ at a particular point of R_n and then defines them throughout all of R_n by parallel displacement. Clearly, any element in E has the form $u = (u, a)a$, where a is the unit vector, while the metrics μ_i and the mappings φ and φ_i^+ have the forms

$$\mu_i(u, v) = \sigma_i(u, a)\overline{(v, a)},$$
$$\varphi h = (h, f)a,$$
$$\varphi_i^+(u) = (u, a)\sigma_i f,$$

where f is the channel element (Chapter IV, §2).

We introduce into consideration the vector $\sigma = \sigma_i e_i$ and assume that the e_i form an orthonormal basis of R_n in a Cartesian coordinate system ($g_{ij} = \delta_{ij}$ and the difference between upper and lower indices is immaterial). From (1) we have

$$2\Gamma_{k,ij} = \sigma_k f_i f_j - \sigma_i f_j f_k - \sigma_j f_k f_i. \tag{2}$$

If $(x, f) = (x, \sigma) = 0$, (2) implies

$$\Gamma_{k,ij} x_k = 0. \tag{3}$$

Suppose f and σ are not colinear and let π denote the two-dimensional space in R_n containing f and σ. From (2) and (3) we see that an "open" Euclidean space that is joined with the external world by only one connecting channel decomposes into a two-dimensional (or one-dimensional, if f and σ are colinear) open space L_2/R_2 and a space R_{n-2}/R_{n-2} that is isolated from the external world (dim E = 0). The fact that the description of L_n in the case dim $E = r$ $(0 < r < \infty)$ requires only a certain subspace of dimension not greater than $r(r+3)/2$ is proved analgously. We direct the x_1 axis in R_2 along the vector f. Then the matrices $\Gamma_{1,ij}$ and $\Gamma_{2,ij}$ take the forms

$$2\Gamma_1 = \begin{Vmatrix} -K_1, & -K_2 \\ K_2, & 0 \end{Vmatrix}, \quad 2\Gamma_2 = \begin{Vmatrix} -K_2, & 0 \\ 0, & 0 \end{Vmatrix},$$

where $K_1 = (\sigma, f)|f|$ and $K_2 = |\sigma \times f| \cdot |f|$ are invariants of the space L_2/V_2. It is easily seen that the open spaces L_2'/R_2' and L_2''/R_2'' are isometric if and only if $K_1' = K_1''$ and $K_2' = K_2''$.

We consider the associated open system equations (3.58), (3.59) on the rays $\dot{x}^i = \xi_i$ of R_n and find their solutions of the forms $ue^{\lambda t}$, $he^{\lambda t}$, $ve^{\lambda t}$. Then, analogously to (4.3), for the input to output mapping we obtain the function

$$S(\xi, \lambda) = I + \varphi(\xi_i \Gamma_i + \lambda I)^{-1} (\xi_i \varphi_i^+),$$

which will be called the *characteristic function of an open space*. This function is easily calculated in the case $L_2/R_2 = \pi$:

$$S(\xi, \lambda) = \frac{4\lambda^2 + 2(\sigma, \xi)|f|^2 \lambda + |\sigma \times f|^2 (h, \xi)^2}{4\lambda^2 - 2(\sigma, \xi)|f|^2 \lambda + |\sigma \times f|^2 (h, \xi)^2}.$$

Clearly, if we know $S(\xi, \lambda)$ for $\xi = f/|f|$, we can find the invariants K_1 and K_2 and recover L_2/R_2 to within an isometry.

Mechanical analogy. Consider a vertical spring with a fixed lower end and a body of mass m attached to its upper end. We join to this body a horizontal semi-infinite string and assume that there is traveling on it both an incident wave in the direction of the body and a reflected wave:

$$y(x, t) = A\,[e^{iw(t+x/c)} - S(w)e^{iw(t-x/c)}], \qquad (0 \leqslant x < \infty),$$

where $S(w)$ is the reflection coefficient, an explicit expression for which is readily found:

$$S(w) = \frac{w^2 + iTc^{-1}w - km^{-1}}{w^2 - iTc^{-1}w - km^{-1}},$$

(k is the spring constant and T is the tension of the string).

If one connects the mechanical and geometrical parameters by the relations

$$\frac{k}{m} = \tfrac{1}{4}(f, \xi)^2 (\sigma \times f)^2, \qquad \lambda = -iw,$$

$$\frac{T}{c} = -\tfrac{1}{2}(\sigma, \xi) f^2,$$

one obtains an interesting coincidence of the reflection coefficient $S(w)$ with the characteristic function $S(\xi, \lambda)$ of the open geometry L_2/V_2.

CHAPTER IV

THE CHARACTERISTIC OPERATOR FUNCTION

§ 1. THE CHARACTERISTIC OPERATOR FUNCTION OF A VECTOR LOCAL COLLIGATION

1. Let $\vec{X} = (\vec{A}, H, \varphi, E, \vec{\mu})$ be a vector local colligation. We consider equations (3.38), (3.39) of the associated open system on any ray $x^k = \xi^k t$ $(k = 1, 2, \ldots, n)$ and choose a special input of the form $u(t) = u(0)\, e^{i\lambda t}$ $(0 \leqslant t < \infty)$. If one seeks a particular solution in the form $h(t) = h(0)\, e^{i\lambda t}$, $v(t) = v(0)\, e^{i\lambda t}$, the corresponding open system equations are written as follows:

$$(\vec{A} \cdot \vec{\xi} - \lambda I)\, h = (\vec{\varphi}^{+} \cdot \vec{\xi})\, u \tag{4.1}$$

$$v = u - i\varphi h, \tag{4.2}$$

where $h = h(0)$, $u = u(0)$ and $v = v(0)$ are constant amplitudes and

$$\vec{A} \cdot \vec{\xi} = \sum_{k=1}^{n} A_k \xi^k, \quad \vec{\varphi}^{+} \cdot \vec{\xi} = \sum_{k=1}^{n} \varphi_k^{+} \xi^k .$$

From equations (4.1) and (4.2) it follows that the input to output mapping $v = S(\vec{\xi}, \lambda) u$ is determined by means of the operator

$$S(\vec{\xi}, \lambda) = I - i\varphi(\vec{A} \cdot \vec{\xi} - \lambda I)^{-1} \quad (\vec{\varphi}^{+} \cdot \vec{\xi}). \tag{4.3}$$

Definition. The operator function S of the variables λ and $\vec{\xi} = (\xi^1, \ldots, \xi^n)$ given by formula (4.3) will be called the *characteristic operator function* (c.o.f.) text dependent of the vector local colligation $\vec{X} = (\vec{A}, H, \varphi, E, \vec{\mu})$.

In particular, when $n = 1$ and $\xi^1 = 1$ we obtain from formula (4.3) the expression

$$S(\lambda) = I - i\varphi(A - \lambda I)^{-1} \varphi^{+} \tag{4.4}$$

for the c.o.f. of a scalar local colligation $X = (A, H, \varphi, E, \mu)$ containing only one internal operator A.

We note that the c.o.f. $S(\vec{\xi}, \lambda)$ of a vector colligation $\vec{X}$ is for each fixed vector $\vec{\xi} = (\xi^1, \ldots, \xi^n)$ the c.o.f. of the scalar colligation

$$\vec{X} \cdot \vec{\xi} = (\vec{A} \cdot \vec{\xi}, H, \varphi, E, \vec{\mu} \cdot \vec{\xi}). \tag{4.5}$$

2. If the complex number λ is a regular point of the operator $\vec{A} \cdot \vec{\xi}$, equations (4.1) and (4.2) determine the pair of mappings

$$R(\vec{\xi}, \lambda) = (\vec{A} \cdot \vec{\xi} - \lambda I)^{-1} \vec{\varphi}^{+} \cdot \vec{\xi}, \tag{4.6}$$

$$S(\vec{\xi}, \lambda) = I - i\varphi R(\vec{\xi}, \lambda) \tag{4.7}$$

of E into H and E respectively. In analogy with the definition of an associated system given in §1 of Chapter III this pair of mappings will be called a *special system* associated with the vector colligation $\vec{X}$.

The proof of the following theorem is analogous to that of Theorem 3.4.

Theorem 4.1. *If a vector local colligation* $\vec{X} = (\vec{A}, H, \varphi, E, \vec{\mu})$ *is an E-product* $\vec{X} = \vec{X}^{(1)} \vee \cdots \vee \vec{X}^{(m)}$ *of vector local colligations* $(\vec{X}^{(k)} = (\vec{A}^{(k)}, H^{(k)}, \varphi^{k}, E, \vec{\mu}))$, *then the special associated system* $F(\vec{\xi}, \lambda)$ *is a coupling* $F(\vec{\xi}, \lambda) = F^{(1)}(\vec{\xi}, \lambda) \vee \cdots \vee F^{(m)}(\vec{\xi}, \lambda)$ *of the special systems* $F^{(k)}(\vec{\xi}, \lambda)$ *associated* *with the colligations* $\vec{X}^{(k)}$:

$$R(\vec{\xi}, \lambda) = R^1(\vec{\xi}, \lambda) + R^2(\vec{\xi}, \lambda) S^1(\vec{\xi}, \lambda) + \cdots + \\ + R^m(\vec{\xi}, \lambda) S^{m-1}(\vec{\xi}, \lambda) \cdots S^1(\vec{\xi}, \lambda), \tag{4.8}$$

$$S(\vec{\xi}, \lambda) = S^m(\vec{\xi}, \lambda) \cdot \ldots \cdot S^1(\vec{\xi}, \lambda). \tag{4.9}$$

Corollary. If $H^{(0)} = H \supset H^{(1)} \supset \cdots \supset H^{(m-1)} \supset H^{(m)} = 0$ *is a chain of subspaces that are invariant under all of the operators* $A_1, \ldots, A_n$, *the c.o.f.* $S(\vec{\xi}, \lambda)$ *can be represented in the form of the product* (4.9), *in which* $S^{(k)}(\vec{\xi}, \lambda)$ *is the c.o.f. of the colligation* $\vec{X}^{(k)} = P^{(k)}\vec{X}$, *where* $P^{(k)}$ *is the orthogonal projection onto the subspace* $H^{(k-1)} \ominus H^{(k)}$ $(k = 1, 2, \ldots, m)$.

§ 2. OPERATOR COMPLEXES

For the purpose of calculating the c.o.f.'s in those cases when $\dim E < \infty$, it is convenient to deal with a so-called operator complex, which is obtained from an operator colligation by choosing an orthonormal basis $a_1, a_2, \ldots, a_m$ in the Hilbert space E.

As we have previously noted, the metrics $\mu_k(u, v)$ can be represented in the form

$$\mu_k(u, v) = (\sigma_k u, v), \tag{4.10}$$

where $(\cdot, \cdot)$ denotes the scalar product in E and the σ_k $(k = 1, 2, \ldots, n)$ are Hermitian operators. As a result, the mapping φ_k^+ can be obtained by noting that

$$\mu_k(\varphi h, v) = (\sigma_k \varphi h, v)_E = (h, \varphi^* \sigma_k v). \tag{4.11}$$

Here φ^* is the adjoint of the operator φ. Consequently,

$$\varphi_k^+ = \varphi^* \sigma_k. \tag{4.12}$$

The local colligation condition can be written in the form

$$\varphi^* \sigma_k \varphi = 2 \operatorname{Im} A_k. \tag{4.13}$$

The elements

$$g_\alpha = \varphi^* a_\alpha \quad (\alpha = 1, 2, \ldots, m) \tag{4.14}$$

of H will be called the *channel elements* while the image $\varphi^* E$ will be called the *channel space* of the vector local colligation $\vec{X}$.

We now write the local colligation condition in terms of the channel elements. Since

$$\varphi h = \sum_{\alpha=1}^{m} (\varphi h, a_\alpha) a_\alpha = \sum_{\alpha=1}^{m} (h, g_\alpha) a_\alpha, \tag{4.15}$$

we have

$$\varphi^* \sigma_k \varphi = \sum_{\alpha=1}^{m} (h, g_\alpha) \sigma_k a_\alpha = \sum_{\alpha=1}^{m} (h, g_\alpha) \sigma_{k,\alpha\beta} g_\beta, \tag{4.16}$$

where

$$\sigma_{k,\alpha\beta} = (\sigma_k a_\alpha, a_\beta). \tag{4.17}$$

Consequently, the local operator colligation condition takes the form

$$\sum_{\alpha,\beta} (\cdot, g_\alpha) \sigma_{k,\alpha\beta} g_\beta = 2 \operatorname{Im} A_k. \tag{4.18}$$

Definition. By an *operator complex* will be meant an ordered set of the form

$$(H, A_1, \ldots, A_n; g_1, \ldots, g_m, \sigma_1, \ldots, \sigma_n), \tag{4.19}$$

where the A_k $(k = 1, 2, \ldots, n)$ are bounded linear operators in H, the g_k $(k = 1, 2, \ldots, m)$ are elements of H and the σ_k $(k = 1, 2, \ldots, n)$ are Hermitian matrices, if condition (4.18) is satisfied.

If an operator complex and an orthonormal basis a_α in E are given, the corresponding colligation can be constructed with the use of relations (4.10) and (4.15).

Let us find the matrix corresponding to the c.o.f. $S(\vec{\xi}, \lambda)$ relative to a basis $\{a_\alpha\}$. To this end we consider the matrix

$$S_{\alpha\beta} = (Sa_\alpha, a_\beta) = I - i(\varphi(\vec{A} \cdot \vec{\xi} - \lambda I)^{-1} \varphi^* (\vec{\sigma}\vec{\xi})\, a_\alpha, a_\beta). \quad (4.20)$$

Since the operators $\varphi(\vec{A} \cdot \vec{\xi} - \lambda I)^{-1} \varphi^*$ and $\vec{\sigma} \cdot \vec{\xi}$ act in the space E, to their product corresponds a product of matrices in reverse order. Hence for the characteristic matrix function (c.m.f.) we obtain[1]

$$S(\vec{\xi}, \lambda) = I - i(\vec{\sigma} \cdot \vec{\xi}) \, \| (\vec{A} \cdot \vec{\xi} - \lambda I)^{-1} g_\alpha, g_\beta) \|, \quad (4.21)$$

where $\vec{\sigma} \cdot \vec{\xi}$ is a matrix of the form $\sum_{k=1}^{n} \sigma_k \xi^k$. If, in particular, $n = 1$, $\xi = 1$ and $\sigma = J$ $(J^2 = I, J = J^*)$, a complex has the form

$$X = (A, H; g_1, \ldots, g_r, J), \quad (4.22)$$

while its c.m.f. is

$$S(\lambda) = I - iJ \, \| ((A - \lambda I)^{-1} g_\alpha, g_\beta) \| . \quad (4.23)$$

We note that if a set of metrics μ_k is nondegenerate, the intersection of the kernels of the operators (matrices) σ_k $(k = 1, 2, \ldots, n)$ is equal to zero:

$$\bigcap_{k=1}^{n} \operatorname{Ker} \sigma_k = 0 . \quad (4.24)$$

[1] In the sequel the same notation will be used for operators and their matrix representations.

Definition. A set of metrics $\mu_1, \ldots, \mu_n$ is said to be *completely nondegenerate* if there exist real numbers $\xi_0^1, \ldots, \xi_0^n$ such that the operator $\sum_{k=1}^{n} \sigma_k \xi_0^k$ is nonsingular.

It is obvious that a completely nondegenerate set of metrics is nondegenerate. Simple examples, which the reader can provide for himself, show that there exist nondegenerate sets of metrics that are not completely nondegenerate.

Remark. A Hilbert space H with a set K of metrics assigned on it can be included in an inductor $(H, \varphi, E, \mu_1, \ldots, \mu_n)$ with a completely nondegenerate set of metrics $\mu_1, \ldots, \mu_n$.

For a proof we consider a Hilbert space H on which a single metric m_k has been assigned. According to Theorem 1.1 of Chapter I a Hilbert space H with a metric assigned on it can be included in a monometric inductor

$$(H, \varphi_k, E_k, \mu_k), \tag{4.25}$$

in which μ_k is a nondegenerate metric.

We form a polymetric inductor

$$(H, \varphi_1 \dot{+} \varphi_2 \dot{+} \cdots \dot{+} \varphi_n, E_1 \dot{+} E_2 \dot{+} \cdots \dot{+} E_n, \tilde{\mu}_1, \ldots, \tilde{\mu}_n), \tag{4.26}$$

where

$$\tilde{\mu}_k = \begin{cases} 0 & \text{on } E_j, \quad \text{if} \quad k \neq j; \\ \mu_k & \text{on } E_k. \end{cases} \tag{4.27}$$

The metric $\tilde{\mu} = \tilde{\mu}_1 + \cdots + \tilde{\mu}_n$ is clearly nondegenerate since the equality

$$\mu_1(u_{0_1}, v_1) + \mu_2(u_{0_2}, v_2) + \cdots + \mu_n(u_{0_n}, v_n) \equiv 0,$$

where $u_0 = \{u_{0_1}, u_{0_2}, \ldots, u_{0_n}\}$ is a fixed element of the space $E_1 \dot{+} E_2 \dot{+} \cdots \dot{+} E_n$ while $v = \{v_1, v_2, \ldots, v_n\}$ is an arbitrary element of this space, implies by virtue of the nondegeneracy of each μ_k on E_k that $u_0 = 0$.

§3. A THEOREM ON THE UNITARY EQUIVALENCE OF LOCAL COLLIGATIONS

Definition. Two vector local colligations $\vec{X} = (\vec{A}, H, \varphi, E, \vec{\mu})$ and $\vec{X}' = (\vec{A}', H', \varphi', E, \vec{\mu})$ are said to be *unitarily equivalent* if there exists an isometric mapping U of H onto H' such that

$$A'_k = UA_kU^{-1}, \tag{4.28}$$

$$\varphi' = \varphi U^{-1}. \tag{4.29}$$

It is easily seen that if two colligations $\vec{X}$ and $\vec{X}'$ are unitarily equivalent, their c.o.f.'s coincide. In fact, by virtue of (4.28) and (4.29) we have

$$S'(\vec{\xi}, \lambda) = I - i\varphi U^{-1}((U\vec{A}U^{-1})\vec{\xi} - \lambda I)^{-1} U\vec{\varphi}^{+}\vec{\xi} = I - i\varphi(\vec{A}\vec{\xi} - \lambda I)^{-1}\vec{\varphi}^{+}\vec{\xi} = S(\vec{\xi}, \lambda).$$

The converse assertion holds under certain additional restrictions. In order to obtain the corresponding result we need some preliminary definitions and lemmas, which will also be used in the sequel.

A colligation $\vec{X} = (\vec{A}, H, \varphi, E, \vec{\mu})$, where the A_k $(k = 1, 2, \ldots, n)$ are pairwise commuting operators, will be said to be *simple* if

$$\{A_1^{k_1}A_2^{k_2} \cdots A_n^{k_n}\varphi^* E\} = H, \tag{4.30}$$

where $\{A_1^{k_1}A_2^{k_2} \cdots A_n^{k_n}\varphi^* E\}$ is the closed linear span of all elements of the form $A_1^{k_1}A_2^{k_2} \cdots A_n^{k_n}\varphi^* u$, in which $k_1, \ldots, k_n$ are nonnegative integers and $u \in E$. Since $\varphi^* a_\alpha = g_\alpha$ $(\alpha = 1, 2, \ldots, m)$, condition (4.30) is equivalent to the condition

$$\{A_1^{k_1} \cdots A_n^{k_n} g_\alpha\} = H. \tag{4.31}$$

If a colligation $\vec{X}$ with pairwise commuting operators A_k $(k = 1, 2, \ldots, n)$ is not simple, the space H can be represented in the form $H = \hat{H} \oplus H_0$, where $\hat{H} = \{A_1^{k_1} \cdots A_n^{k_n}\varphi^* E\}$, in which case $\hat{H}$ and H_0 are invariant subspaces under the operators A_k $(k = 1, 2, \ldots, n)$ such that $\vec{X}$ is simple on $\hat{H}$ and has the following form on H_0:

$$\vec{X}_0 = (\vec{A}, H_0, \varphi_0 \equiv 0, E, \vec{\mu}). \tag{4.32}$$

In fact, the invariance of $\hat{H}$ under the operators A_k $(k = 1, 2, \ldots, n)$ is obvious. Further, if $h_0 \in H_0$,

$$\mu_k(\varphi h_0, u) = (h_0, \varphi^* \sigma_k u) = 0 \quad (k = 1, 2, \ldots, n; \; u \in E) \tag{4.33}$$

and consequently $\varphi h_0 \in \bigcap_{k=1}^{n} \mathrm{Rad}\,\mu_k = 0$. From the local colligation condition we have $(A_k - A_k^*)\, h_0 = i\varphi_k^+ \varphi h_0 = 0$ and hence $A_k h_0 = A_k^* h_0$ $(h_0 \in H_0)$. It follows that $(A_k h_0, \hat{h}) = (A_k^* h_0, \hat{h}) = (h_0, A_k \hat{h}) = 0$ $(\hat{h} \in \hat{H})$ and thus $A_k h_0 \in H_0$.

The subspace $H_0 = H \ominus \hat{H}$ is called the *complementary component* of $\vec{X}$ while $\hat{H}$ is called the *principal component* of $\vec{X}$.

Lemma 4.1. *If A is a bounded linear operator in H*

$$e^{itA} = -\frac{1}{2\pi i} \oint e^{i\lambda t} (A - \lambda I)^{-1}\, d\lambda, \tag{4.34}$$

where the contour encloses the spectrum of the operator A.

(For a proof see [11].)

Consider a scalar product in R_n of the form

$$v(x, y) = (h_{1_x}, h_{2_y}) \quad (x, y \in R_n), \tag{4.35}$$

where

$$h_{\nu_x} = e^{i \sum_{k=1}^{n} A_k x^k} h_\nu \quad (\nu = 1, 2), \tag{4.36}$$

in which the h_ν are fixed elements of H.

Let

$$W_k(x, y) = -\frac{\partial v(x+\tau, y+\tau)}{\partial \tau^k}\bigg|_{\tau=0}. \tag{4.37}$$

Lemma 4.2.

$$W_k(x, y) = (2\,\mathrm{Im}\,A_k h_{1_x}, h_{2_y}) = \sum_{\alpha,\beta=1}^{m} (h_{1_x}, g_\alpha)\,\sigma_{k,\alpha\beta}\,(g_\beta, h_{2_y}) \tag{4.38}$$

where the g_α *are channel elements* (see (4.14)).

In fact,

$$-\frac{\partial}{\partial \tau^k} v(x+\tau, y+\tau)\,|_{\tau=0} = -\frac{\partial}{\partial \tau_k}\left(e^{i\sum_{j=1}^{n} A_j(x^j+\tau^j)} h_{1_x}, e^{i\sum_{j=1}^{n} A_j(y^j+\tau^j)} h_{2_y}\right) =$$

$$\left(\left(\frac{A_k - A_k^*}{i}\right) e^{i\sum_{j=1}^{n} A_j x^j} h_1, e^{i\sum A_k y^k} h_2\right) = (2\,\mathrm{Im}\,A_k h_{1_x}, h_{2_y}).$$

Lemma 4.3. *A local colligation (complex) with pairwise commuting operators* A_k $(k = 1, 2, \ldots, n)$ *will be simple if and only if*

$$\{e^{i\sum A_k x^k} g_\alpha\} = H \quad (x \in R_n,\ \alpha = 1, 2, \ldots, m). \tag{4.39}$$

A proof of the lemma follows directly from the expansion

$$e^{i\sum A_k x^k} g_\alpha = \sum_{k_1 \cdots k_n} \frac{i^{k_1+k_2+\cdots+k_n}}{k_1!\,k_2!\cdots k_n!} (x^1)^{k_1} (x^2)^{k_2} \cdots (x^n)^{k_n} A_1^{k_1} A_2^{k_2} \cdots A_n^{k_n} g_\alpha.$$

Lemma 4.4. *Two colligations* $\vec{X}$ *and* $\vec{X}'$ *are unitarily equivalent if*

1) $A_k A_j = A_j A_k$, $A_k' A_j' = A_j' A_k'$, (4.40)

2) *they are simple*, (4.41)

$$3)\ \left(e^{i\sum A_k x^k} g_\alpha,\ e^{i\sum A_k y^k} g_\beta\right) = \left(e^{i\sum A'_k x^k} g'_\alpha,\ e^{i\sum A'_k y^k} g'_\beta\right). \tag{4.42}$$

Proof. We first define the required mapping U on the linear manifold spanned by elements of the form $\sum_{j,\alpha} c_{j,\alpha} h_\alpha(x_j)$ by setting

$$U\left(\sum_{j,\alpha} c_{j,\alpha} h_\alpha(x_j)\right) = \sum_{j,\alpha} c_{j,\alpha} h'_\alpha(x_j) \tag{4.43}$$

where

$$\begin{aligned} h_\alpha(x) &= e^{i\sum_k A_k x^k} g_\alpha, \\ h'_\alpha(x) &= e^{i\sum_k A'_k x^k} g'_\alpha . \end{aligned} \tag{4.44}$$

This mapping is a single-valued isometric operator by virtue of condition (4.42) and can be extended by continuity onto the space $\left\{\sum c_{j,\alpha} h_\alpha(x_j)\right\}$. By virtue of Lemma 4.3 it is actually defined on all of H and its range coincides with H. It therefore follows from (4.43) that

$$U e^{i\sum A_k x^k} g_\alpha = e^{i\sum A'_k x^k} g'_\alpha . \tag{4.45}$$

Hence

$$g'_\alpha = U g_\alpha \tag{4.46}$$

and

$$U A_k h_\alpha(x) = A'_k h'_\alpha(x). \tag{4.47}$$

Since $h'_\alpha(x) = U h_\alpha(x)$, we have $U A_k = A'_k U$ on $\left\{\sum c_{j,\alpha} h_\alpha(x_j)\right\}$ and hence on all of H.

Further, by virtue of (4.15) and (4.46)

$$\varphi h = \sum_{\alpha} (h, g_{\alpha}) a_{\alpha} = \sum_{\alpha} (h, U^{-1} g'_{\alpha}) a_{\alpha} = \sum_{\alpha} (Uh, g'_{\alpha}) a_{\alpha} = \varphi' U h,$$

and the lemma is proved.

Theorem 4.2. *The principal components of two vector local colligations* $\vec{X} = (A, H, \varphi, E, \vec{\mu})$ *and* $\vec{X}' = (\vec{A}', H', \varphi', E, \vec{\mu})$ *are unitarily equivalent if*

1) $A_k A_j = A_j A_k \quad (k, j = 1, 2, \ldots, n),$ (4.48)

2) $A'_k A'_j = A'_j A'_k \quad (k, j = 1, 2, \ldots, n),$ (4.49)

3) *the system of metrics* $\vec{\mu} = (\mu_1, \ldots, \mu_n)$ *is completely nondegenerate and*

4) *the c.o.f.'s of* $\vec{X}$ *and* $\vec{X}'$ *coincide at regular points:*

$$S(\vec{\xi}, \lambda) = S'(\vec{\xi}, \lambda). \tag{4.50}$$

Proof. It can be assumed without loss of generality that the colligations $\vec{X}$ and $\vec{X}'$ are simple. Since the system of metrics μ_k $(k = 1, 2, \ldots, n)$ is completely nondegenerate, there exists a point $\xi_0 \in R_n$ such that the metric $\vec{\mu} \cdot \vec{\xi} = \sum_{k=1}^{n} \mu_k \xi^k$ is nondegenerate in a neighborhood of it. From condition (4.50) and formula (4.21) for the c.o.f. we see that the following equality then holds in this neighborhood:

$$((\vec{A} \cdot \vec{\xi} - \lambda I)^{-1} g_{\alpha}, g_{\beta}) = ((\vec{A}' \cdot \vec{\xi} - \lambda I)^{-1} g'_{\alpha}, g'_{\beta}). \tag{4.51}$$

With the use of Lemma 4.1 we get

$$(h_{\gamma}(\vec{\xi}), g_{\delta}) = -\frac{1}{2\pi i} \oint e^{i\lambda} ((\vec{A} \cdot \vec{\xi} - \lambda I)^{-1} g_{\gamma}, g_{\delta}) \, d\lambda, \tag{4.52}$$

where

$$h_\gamma(x) = e^{i\sum_{k=1}^{n} A_k x^k} g_\gamma. \tag{4.53}$$

Therefore, from (4.51) and (4.52) we obtain the relation

$$(h_\gamma(\xi),\ g_\delta) = (h'_\gamma(\xi),\ g'_\delta), \tag{4.54}$$

which holds in all of R_n by virtue of its analyticity relative to the parameters $\xi^1, \xi^2, \ldots, \xi^n$.

It follows from (4.38) and (4.54) that

$$W_{k,\gamma\delta}(x,\ y) = W'_{k,\gamma\delta}(x,\ y), \tag{4.55}$$

where

$$W_{k,\gamma\delta}(x,\ y) = \sum_{\alpha,\beta} (h_\gamma(x),\ g_\alpha)\, \sigma_{k,\alpha\beta}\, \overline{(h_\delta(y),\ g_\beta)}. \tag{4.56}$$

In addition, from (4.38) we obtain the relation

$$v_{\gamma\delta}(x,\ y) - v_{\gamma\delta}(x-y,\ 0) = \int_0^{-y} W_{k,\gamma\delta}(x+\tau,\ y+\tau)\, d\tau^k, \tag{4.57}$$

where

$$v_{\gamma\delta}(x,\ y) = (h_\gamma(x),\ h_\delta(y)), \tag{4.58}$$

while the integral in (4.57) is taken along any path joining the points 0 and $-y$ of R_n.

Further, since

$$v_{\gamma\delta}(x-y,\ 0) = (h_\gamma(x-y),\ g_\delta), \tag{4.59}$$

it follows from (4.54), (4.55) and (4.57) that

$$(h_\gamma(x),\ h_\delta(y)) = (h'_\gamma(x),\ h'_\delta(y)) \quad (\gamma,\ \delta = 1,\ 2,\ \ldots,\ m). \tag{4.60}$$

But this result together with Lemma 4.4 implies the assertion of the theorem.

§ 4. QUASI-HERMITIAN COLLIGATIONS

A scalar local colligation $X = (A,\ H,\ \varphi,\ E,\ \mu)$ is said to be *quasi-Hermitian* if the operator φ is completely continuous.

Let J be an involution in $E\ (J = J^*,\ \ J^2 = I)$. An operator function $S(\lambda)$ whose values are bounded linear operators in E will be said to belong to the *class* Ω_J if

a) it is holomorphic in the complement G_s in the extended complex plane of a bounded set all of whose limit points are real,

b) $\lim\limits_{\lambda\to\infty} \| S(\lambda) - I \| = 0,$

c) $S^*(\lambda) JS(\lambda) - J \geqslant 0 \quad (\operatorname{Im}\lambda > 0,\ \lambda \in G_s),$

d) $S^*(\lambda) JS(\lambda) - J = 0 \quad (\operatorname{Im}\lambda = 0,\ \lambda \in G_s),$

e) all of the operators $S(\lambda) - I\ (\lambda \in G_s)$ are completely continuous.

Theorem 4.3. *In order for a given operator function* $S(\lambda)$ *to be the c.o.f. of a simple quasi-Hermitian colligation with a* J*-metric* $\mu(u,\ v) = (Ju,\ v)\ (u,\ v \in E)$ *it is necessary and sufficient that it belong to the class* Ω_J.

A proof can be found in [5].

The subspace $G = \overline{(2\operatorname{Im} A)H}$, which is the closure of the range of the operator $2\operatorname{Im} A = \dfrac{A - A^*}{i}$, is called the *non-Hermitian subspace* of A. If the closed linear span $\{A^n G\}_{n=0}^{\infty}$ coincides with H, the operator A is said to be *completely non-self-adjoint*. Since the subspace $\overline{\varphi^* E}$ contains the non-Hermitian subspace G by virtue of (3.37), a local colligation X whose internal operator A is completely non-self-adjoint is simple.

The converse assertion is not true, since any bounded operator can be included in a simple colligation [4].

If a colligation $X = (A,\ H,\ \varphi,\ E,\ \mu)$ is quasi-Hermitian, the operator $2\operatorname{Im} A$ is completely continuous by virtue of the condition $2\operatorname{Im} A = \varphi^+\varphi$.

Theorem 4.4. *An operator A whose imaginary part* $\operatorname{Im} A$ *is completely continuous has the following properties:*

1) *any nonreal point λ is either a regular point or a pole of the resolvent of A;*

2) *every nonreal point λ_0 of the spectrum of A is an eigenvalue with a finite-dimensional generalized eigenspace;*

3) *the limit points of the nonreal spectrum can lie only on the real axis.*

A proof of this theorem is given in [8].

An operator A is said to be *dissipative* if $\operatorname{Im} A \geqslant 0$. A colligation $X = (A, H, \varphi, E, \mu)$ is said to be *dissipative* if μ is a J-metric and $J = I$. In this case $\mu(u, v) = (u, v)_E$, where $(u, v)_E$ is the Hilbert scalar product.

The internal operator A of a dissipative colligation is clearly dissipative. Every dissipative operator can be included in a dissipative colligation [4]. The channel subspace of a dissipative colligation coincides with the non-Hermitian subspace [4]. A dissipative colligation is simple if and only if its internal operator A is completely non-self-adjoint.

We note in addition that the spectrum of a dissipative operator is contained in the closed upper halfplane.

According to (4.18) and (4.19) there corresponds to an operator colligation $X = (A, H, \varphi, E, \mu)$ $(\dim E = r < \infty)$ with a J metric $\mu(u, v) = (Ju, v)_E$ a complex $X = (A, H, g_1, \dots, g_r, J)$ such that

$$2\operatorname{Im} A = \sum_{\alpha, \beta=1}^{r} (\cdot, g_\alpha) J_{\alpha\beta} g_\beta \quad (J_{\alpha\beta} = (Ja_\alpha, a_\beta)). \tag{4.61}$$

We choose a basis a_α so that J is a diagonal matrix. Then

$$2\operatorname{Im} A = \sum_{\alpha=1}^{r} \varepsilon_\alpha (\cdot, g_\alpha) g_\alpha \quad (\varepsilon_\alpha = \pm 1). \tag{4.62}$$

For an arbitrary dissipative complex $X = (A, H, g_1, \dots, g_r, I)$ condition (4.62) takes the form

$$2\operatorname{Im} A = \sum_{\alpha=1}^{r} (\cdot, g_\alpha) g_\alpha . \tag{4.63}$$

If h_k $(k = 1, 2, \ldots)$ is an orthonormal basis of H, relation (4.63) implies

$$\operatorname{Tr}(2\operatorname{Im} A) = \sum_{k=1}^{\infty} (2\operatorname{Im} A h_k, h_k) = \sum_{\alpha=1}^{r} \sum_{k=1}^{\infty} (h_k, g_\alpha)(g_\alpha, h_k) \tag{4.64}$$

and since $\sum_{k=1}^{\infty} |(g_\alpha, h_k)|^2 = \| g_\alpha \|^2$, the trace of the operator $2\operatorname{Im} A$ is given by the relation

$$\operatorname{Tr}(2\operatorname{Im} A) = \sum_{\alpha=1}^{r} \| g_\alpha \|^2. \tag{4.65}$$

Using (4.65) and the fact that under a multiplication $X = X_2 \vee X_1$ of operator complexes their channel elements are added ($g_\alpha = g_{1\alpha} \oplus g_{2\alpha}$) by virtue of (1.14), we obtain for the corresponding dissipative operators the relation

$$\operatorname{Tr} \operatorname{Im} A = \operatorname{Tr} \operatorname{Im} A_1 + \operatorname{Tr} \operatorname{Im} A_2 . \tag{4.66}$$

We also note that if H_1 is an invariant subspace of a dissipative operator A,

$$Sp \operatorname{Im} A_1 \leqslant Sp \operatorname{Im} A, \tag{4.67}$$

where A_1 is the restriction of A to H_1.

If H' is a finite-dimensional invariant subspace of a dissipative operator A and $\lambda_1, \lambda_2, \ldots, \lambda_n$ constitute the nonreal spectrum of A in H' then, as can easily be seen,

$$\operatorname{Tr} A' = \sum_{k=1}^{n} \operatorname{Im} \lambda_k . \tag{4.68}$$

This result together with (4.67) implies the inequality

$$\sum_{k=1}^{\infty} \operatorname{Im} \lambda_k \leqslant \operatorname{Tr} \operatorname{Im} A, \tag{4.69}$$

where $\lambda_1, \lambda_2, \ldots, \lambda_n, \ldots$ is the nonreal spectrum of A.

§ 5. ELEMENTARY DISSIPATIVE COMPLEXES

1. Let us agree to call a complex $X_0 = (A_0, H_0, g_1^0, \ldots, g_r^0, I)$ *elementary* if $\dim H_0 = 1$. In this case the operator A_0 has the form

$$Ah = \lambda_0 h \quad (h \in H), \tag{4.70}$$

where λ_0 is a complex number such that $\operatorname{Im} \lambda_0 \geqslant 0$. Since all of the channel elements belong to a one-dimensional space, they have the form

$$g_\alpha^0 = q_\alpha h_0 \quad (\alpha = 1, 2, \ldots, r), \tag{4.71}$$

where q_α is a number and h_0 is the unit vector of H. It is easily seen that the condition (4.18) for a complex is in the present case equivalent to the equality

$$2\operatorname{Im} \lambda_0 = \sum_{\alpha=1}^{r} |q_\alpha|^2. \tag{4.72}$$

Let us find the c.m.f. $S_0(\lambda)$ of an elementary complex. According to formula (4.23)

$$S(\lambda) = I - i\,\|((A - \lambda I)^{-1} g_\alpha^0, g_\beta^0)\|. \tag{4.73}$$

Since the resolvent $(A_0 - \lambda I)^{-1} h = \frac{h}{\lambda - \lambda_0}$, we get

$$S_0(\lambda) = I - \frac{i}{\lambda - \lambda_0} \| q_\alpha \bar{q}_\beta \|. \tag{4.74}$$

Formula (4.74) can be written in the form

$$S_0(\lambda) = I - \frac{i}{\lambda - \lambda_0} q^* q, \tag{4.75}$$

where $q = \|\bar{q}_1, \bar{q}_2, \ldots, \bar{q}_r\|$ denotes the row matrix composed of the numbers $\bar{q}_\alpha$ $(\alpha = 1, 2, \ldots, r)$. In this notation condition (4.72) takes the form

$$qq^* = 2\operatorname{Im} \lambda_0. \tag{4.76}$$

We note that $r-1$ eigenvalues of a matrix of the form $\| q_\alpha \bar{q}_\beta \|$ are equal to zero while the remaining one is equal to $\sum_{\alpha=1}^{r} | q_\alpha |^2 = 2\operatorname{Im} \lambda_0$. Hence the determinant of an elementary c.o.f. is given by the expression

$$\det S_0(\lambda) = \frac{\lambda - \bar{\lambda}_0}{\lambda - \lambda_0}. \tag{4.77}$$

2. We consider another complex with a special form:

$$Z = (A_0, H_r, h_1, h_2, \ldots, h_r, I), \tag{4.78}$$

where $\dim H_r = r$, the operator A_0 is given on H_r by the equality $A_0 h = \lambda_0 h$, and the channel elements have the form

$$h_\alpha = \sqrt{2\operatorname{Im} \lambda_0}\, e_\alpha \tag{4.79}$$

in which $(e_\alpha e_\beta) = \delta_{\alpha\beta}$ $(\alpha, \beta = 1, 2, \ldots, r)$. It is obvious that the condition for a complex

$$2\operatorname{Im} A_0 h = \sum_{\alpha=1}^{r} (h, h_\alpha) h_\alpha$$

is satisfied in the present case.

The c.m.f. is readily calculated:

$$S_Z(\lambda) = \frac{\lambda - \bar{\lambda}_0}{\lambda - \lambda_0} I \tag{4.80}$$

Consider an arbitrary one-dimensional subspace H_0 of H_r. Since H_0 is invariant relative to A_0, the complex $Z = (P_0^\perp Z) \vee (P_0 Z)$ (see Chapter I, §4), where P_0 and $P_0^\perp$ are the orthogonal projections onto H_0 and $H_r \ominus H_0$. It follows by virtue of (4.9) that

$$S_z(\lambda) = S_0^\perp(\lambda) S_0(\lambda) = S_0(\lambda) S_0^\perp(\lambda), \tag{4.81}$$

where $S_0^\perp(\lambda)$ and $S_0(\lambda)$ are the c.m.f.'s of the complexes $P_0^\perp Z$ and $P_0 Z$ respectively. We note further that the complex

$$P_0 Z = (A_0, H_0, P_0 h_1, \ldots . P_0 h_r, I) \tag{4.82}$$

is elementary. Its channel elements have the form

$$g_\alpha^0 = P^0 h_\alpha = \sqrt{2\operatorname{Im}\lambda_0}\,(e_\alpha,\ h_0)\,h_0 \tag{4.83}$$

while the numbers q_α in (4.76) are equal to

$$q_\alpha = \sqrt{2\operatorname{Im}\lambda_0}\,(e_\alpha,\ h_0). \tag{4.84}$$

By arbitrarily varying the unit vector h_0 ($h_0 \in H_r$) in (4.84), we can obtain any of the numbers q_α satisfying condition (4.72). This result together with (4.81) implies the following theorem.

Theorem 4.5. *If* $S_0(\lambda)$ *is an elementary c.m.f. of form* (4.74), *the product*

$$\frac{\lambda - \bar{\lambda}_0}{\lambda - \lambda_0} S_0^{-1}(\lambda) = S_0^{\perp}(\lambda) \tag{4.85}$$

is the c.m.f. of some dissipative complex such that $\dim H_0^{\perp} = r - 1$.

Definition. If

$$S(\lambda) = S_1(\lambda)\, S_2(\lambda), \tag{4.86}$$

where $S(\lambda)$, $S_1(\lambda)$ and $S_2(\lambda)$ are matrix functions of class Ω_J, one says that $S_1(\lambda)$ is a *left divisor*, while $S_2(\lambda)$ is a *right divisor*, of $S(\lambda)$.

Theorem 4.5 implies

Theorem 4.6. *A matrix function* $S_0(\lambda)$ *of form* (4.74) *under condition* (4.76) *belongs to the class* Ω_J *and is a divisor (both left and right) of the matrix function* $\frac{\lambda - \bar{\lambda}_0}{\lambda - \lambda_0} I$.

NOTES

§1. An account of the theory of the characteristic operator functions of local colligations can be found in the monograph of M.S. Brodskiĭ [4] and in the survey article of M.S. Brodskiĭ and M.S. Livshits [5].

Characteristic matrix functions of quasiunitary operators were introduced and studied in [28], [34]. A.V. Kužel' [24^{3,4}] investigated characteristic matrix functions for linear operators in pseudounitary spaces.

Characteristic functions for contraction operators were the object of a deep study by B. Sz.-Nagy and S. Foias [40]. The theory of characteristic operator functions for nonunitary operator was developed in terms of operator colligations in [7¹], [6¹], [27²], [7²], [6²]; in particular, V. M. Brodskiĭ [6¹], [6²] sharpened the notion of a metric colligation introduced by M. S. Livshits [27²] by calling a quintuple $[T, H, \phi\ E, K)$ of Hilbert spaces H and E and operators $T(H \overset{T}{\to} H)$, $K(E \overset{K}{\to} E)$, and $\Phi(E \overset{\Phi}{\to} H)$ a *U-colligation* if T is invertible and $TT^* - I = \Phi\Phi^+$, $KK^+ - I = \Phi^+\Phi$. With each *U-colligation* there is associated a characteristic operator function $\theta(\zeta) = (K^+)^{-1}[I - \Phi^+(I\zeta - T)^{-1}\Phi]$, for which a theory analogous to the theory of characteristic operator functions of local colligations is developed. We note that there can be associated with each *U-colligation* a discrete open system given by equations of the forms

$$\psi_{n+1} - T\psi_n = \phi u_n,$$
$$K^+ v_n = u_n + \phi^+ \psi_{n+1}, \tag{1}$$

where the $\psi_n \in H$ $(n = 0, 1, 2, \ldots)$ constitute the internal state and the u_n, v_n constitute the input and output respectively.

As in the case of open systems with continuous time, there exists a *metric conservation principle*, which in the case of discrete time has the form

$$\|\psi_{n+1}\|^2 - \|\psi_n\|^2 = \mu(v_n, v_n) - \mu(u_n, u_n).$$

A coupling of open system with discrete time is defined in exactly the same way as in the continuous case (see (3.27), (3.28)).

A product can be introduced for *U-colligations* (the analog of the E-product for local colligations) as follows: an aggregate $U = (T, H, \phi, E, K)$ is called the *product* of the *U-colligations* U_1 and U_2, where $U_i = (T_i, H_i, \phi_i, E, K_i)$ $(i = 1, 2)$, if

$$T = T_1P_1 + T_2P_2 - \phi_2 K_1{}^{-1}\phi_1^+ T_1 P_1,$$

$$\phi = \phi_1 + \phi_2 K_1^+, \qquad K = K_2K_1, \qquad H = H_1 \oplus H_2.$$

This product operation for *U-colligations* can be used to prove the following theorem.

Let F_i be the open systems respectively associated with the U-colligations $U_i = (T_i, H_i, \phi_i, E, K_i)$ $(i = 1, 2, \ldots, m)$. Then the coupling $F = F_1 \vee F_2 \vee \ldots \vee F_m$ is the open system associated with the product of the U-colligations U_i.

Returning to equations (1), we note that when $\psi_n = \psi\zeta^n$, $u_n = u\zeta^n$ and $v_n = v\zeta^n$, the input to output mapping coincides with the characteristic operator function $\theta(\zeta)$ of the corresponding *U-colligation.*

A characteristic operator function for unbounded operators has been considered by A. V. Kužel' [24[1,2]] and A. V. Straus [38].

§3. V. E. Kacnel'son has given another proof of Theorem 4.2, which is close to the known proof for the case $n = 1$.

In connection with Theorem 4.2 there arises the problem of extending the whole theory of characteristic functions and triangular representations to systems of commuting operators.

Another interesting problem consists in determining how essential are the commutativity conditions

$$A_j A_k = A_k A_j, \quad A'_j A'_k = A'_k A'_j, \qquad (j, k = 1, 2, \ldots, n).$$

in Theorem 4.2.

It is shown in [41] that Theorem 4.2 generalizes to the case when the linear spaces of operators of the forms $\sum_{k=1}^{n} \xi^k A_k$ and $\sum_{k=1}^{n} \xi^k A'_k$ are Lie algebras with the same structure constants, i.e. to the case when conditions (4.48) and (4.49) can be replaced by the more general commutator conditions

$$[A_i, A_j] = \sum_{k=1}^{n} C_{ij}{}^k A_k,$$

$$[A'_i, A'_j] = \sum_{k=1}^{n} C_{ij}{}^k A'_k.$$

CHAPTER V*

TRIANGULAR AND UNIVERSAL MODELS OF LINEAR OPERATORS

§1. OPERATORS OF NON-HERMITIAN RANK ONE

1. The *non-Hermitian rank* of an operator A is the dimension of its non-Hermitian space $G = \overline{2\,(\mathrm{Im}\,A)\,H}$. Consider a local colligation $(H,\ A,\ \varphi,\ E,\ \mu)$ such that $\dim E = 1$ and μ is a J-metric. In this case $J = \pm 1$ and it can be assumed without loss of generality that $J = 1$. The corresponding operator complex has the form

$$X = (A,\ H,\ g,\ 1), \tag{5.1}$$

where g is the channel element and A satisfies the condition

$$2\,(\mathrm{Im}\,A)\,h = (h,\ g)\,g \quad (h \in H). \tag{5.2}$$

Since $\dim E = 1$, the characteristic function in the case under consideration is the scalar function

$$S(\lambda) = 1 - i\,((A - \lambda I)^{-1} g,\ g), \tag{5.3}$$

*In this chapter (except for §4) we present well-known results that are needed for an understanding of Chapters VII–VIII.

which in accordance with Theorem 4.3 satisfies the following conditions:

1) it is holomorphic in the complement in the extended complex plane of the nonreal spectrum of A;

2) $\lim_{\lambda\to\infty} |S(\lambda) - 1| = 0$;

3) $|S(\lambda)| \geqslant 1 \quad (\operatorname{Im}\lambda > 0)$;

4) $|S(\lambda)| = 1 \quad (\operatorname{Im}\lambda = 0, \quad |\lambda| > \|A\|)$.

Theorem 5.1. *The function* $S(\lambda)$ *admits the representation*

$$S(\lambda) = \prod_{k=1}^{N} \frac{\lambda - \bar{\lambda}_k}{\lambda - \lambda_k} e^{-i\int_a^b \frac{d\sigma(t)}{t-\lambda}}, \tag{5.4}$$

where the λ_k $(k = 1, 2, \ldots, N;\ N \leqslant \infty)$ *are the nonreal eigenvalues of the operator* A *and* $\sigma(t)$ *is a bounded nondecreasing function* $(-\infty < a \leqslant \leqslant t \leqslant b < \infty)$. *Each number* λ_k *is repeated as many times as the dimension of the corresponding invariant subspace.*

Proof. Since $\sum_{k=1}^{N} \operatorname{Im}\lambda_k < \infty$ by virtue of (4.69), the product

$$S_0(\lambda) = \prod_{k=1}^{N} \frac{\lambda - \bar{\lambda}_k}{\lambda - \lambda_k} = \prod_{k=1}^{N} \left(1 + \frac{\lambda_k - \bar{\lambda}_k}{\lambda - \lambda_k}\right) \tag{5.5}$$

converges and satisfies conditions 1)–4). It is easily seen that the function $S_1(\lambda) = \frac{S(\lambda)}{S_0(\lambda)}$ also satisfies these conditions and does not have any poles or zeros in the region $\operatorname{Im}\lambda \neq 0$.

If one puts $v(\lambda) = i \ln S_1(\lambda)$, conditions 2) and 3) imply $\operatorname{Im} v(\lambda) \geqslant 0$ $(\operatorname{Im}\lambda \geqslant 0)$ and, in particular, $\operatorname{Im} v(\lambda) = 0$ on the real axis outside the interval $|\lambda| \leqslant \|A\|$. As a result (see [1]), we have the representation

$$v(\lambda) = \int_a^b \frac{d\sigma(\lambda)}{t-\lambda}. \tag{5.6}$$

The representation (5.4) follows from equalities (5.5) and (5.6).

Definition. A countable or finite set of complex numbers $\lambda_1, \lambda_2, \ldots, \lambda_n, \ldots$, satisfying the conditions

$$1)\ \operatorname{Im}\lambda_k > 0 \quad (k = 1, 2, \ldots), \tag{5.7}$$

$$2)\ \sum_{k=1}^{N} \operatorname{Im}\lambda_k < \infty, \tag{5.8}$$

$$3)\ |\lambda_k| < c \quad (k = 1, 2, \ldots), \tag{5.9}$$

will be called a λ*-set*.

From what has been said above it follows that the nonreal spectrum of the operator A forms a λ-set.

Let $\sigma(t) = s$ in the representation (5.4), which then takes the form

$$S(\lambda) = \prod_{k=1}^{N} \frac{\lambda - \bar{\lambda}_k}{\lambda - \lambda_k} e^{-i\int_0^l \frac{ds}{\alpha(s)-\lambda}}, \tag{5.10}$$

where l is the variation of $\sigma(t)$ on [a,b] and $\alpha(s)$ is the inverse of the function $\sigma(t) = s$ which is right-continuous, and such that if t_0 is a point of discontinuity of $\sigma(t)$, the identity $\alpha(s) \equiv t_0$ holds in the interval $\sigma(t_0 - 0) \leqslant s < \sigma(t_0 + 0)$.

2. Let us first consider the case $l = 0$. Then $S(\lambda)$ has the form

$$S(\lambda) = \prod_{k=1}^{N} \frac{\lambda - \bar{\lambda}_k}{\lambda - \lambda_k} \quad (N \leqslant \infty), \tag{5.11}$$

where $\lambda_k = \alpha_k + \frac{i}{2}\beta_k^2$ and, by assumption, $\sum_{k=1}^{N} \beta_k^2 < \infty$ $(-\infty < \alpha_k < \infty,\ \beta_k > 0)$.

We introduce the space l_2 of all functions $f(k)$ on the positive integers $(k = 1, 2, \ldots)$ with scalar product

$$(f_1, f_2) = \sum_{k=1}^{N} f_1(k)\overline{f_2(k)}. \tag{5.12}$$

Further, we define an operator $\dot{A}f$ in l_2:

$$(Af)_k = \left(\alpha_k + i\frac{\beta_k^2}{2}\right) f(k) + i \sum_{j=k+1}^{N} f(j)\beta_j\beta_k \quad (k = 1, 2, \ldots, N) \tag{5.13}$$

or, in matrix notation,

$$\dot{A} = \begin{Vmatrix} \alpha_1 + \frac{i}{2}\beta_1^2 & i\beta_1\beta_2 & \ldots & i\beta_1\beta_k & \ldots \\ 0 & \alpha_2 + \frac{i}{2}\beta_2^2 & \ldots & i\beta_2\beta_k & \ldots \\ \cdots & \cdots & \cdots & \cdots & \cdots \\ \cdots & \cdots & \cdots & \cdots & \cdots \\ 0 & 0 & \ldots & \alpha_k + \frac{i}{2}\beta_k^2 & \ldots \\ \cdots & \cdots & \cdots & \cdots & \cdots \end{Vmatrix}. \tag{5.14}$$

This operator, which will be called a *triangular model*, is bounded. In fact, it is the sum of two bounded operators corresponding to the two terms in (5.13). The boundedness of the first is obvious while the boundedness of the second follows from the inequality

$$\left|\sum_{j=2}^{N} f(j)\beta_j\beta_1\right|^2 + \left|\sum_{j=3}^{N} f(j)\beta_j\beta_2\right|^2 + \cdots \leqslant$$

$$\|f\|^2\left(\beta_1^2 \sum_{j=1}^{N} \beta_j^2 + \beta_2^2 \sum_{j=1}^{N} \beta_j^2 + \cdots\right) \leqslant \|f\|^2\left(\sum_{j=1}^{N} \beta_j^2\right)^2.$$

Going over to $\dot{A}^*$, we have

$$(\dot{A}^* f)_k = \left(\alpha_k - i\frac{\beta_k^2}{2}\right) f(k) - i\sum_{j=1}^{k-1} f(j)\,\beta_j\beta_k \tag{5.15}$$

or, in matrix notation,

$$\dot{A}^* = \left\| \begin{matrix} \alpha_1 - i\frac{\beta_1^2}{2} & 0 & 0 & \cdots & 0 & \cdots \\ -i\beta_2\beta_1 & \alpha_2 - i\frac{\beta_2^2}{2} & 0 & \cdots & 0 & \cdots \\ \cdots & \cdots & \cdots & \cdots & \cdots & \cdots \\ \cdots & \cdots & \cdots & \cdots & \cdots & \cdots \\ -i\beta_k\beta_1 & -i\beta_k\beta_2 & \cdots & \cdots & \alpha_k - i\frac{\beta_k}{2} & \cdots \\ \cdots & \cdots & \cdots & \cdots & \cdots & \cdots \end{matrix} \right\|. \tag{5.16}$$

Therefore

$$\frac{1}{i}(\dot{A} - \dot{A}^*) = \left\| \begin{matrix} \beta_1^2 & \beta_1\beta_2 & \cdots & \beta_1\beta_k & \cdots \\ \beta_2\beta_1 & \beta_2^2 & \cdots & \beta_2\beta_k & \cdots \\ \cdots & \cdots & \cdots & \cdots & \cdots \\ \beta_k\beta_1 & \beta_k\beta_2 & \cdots & \beta_k^2 & \cdots \\ \cdots & \cdots & \cdots & \cdots & \cdots \end{matrix} \right\| \tag{5.17}$$

and hence

$$[2(\mathrm{Im}\,\dot{A})f]_k = \sum_{j=1}^{N} f(j)\,\beta_j\beta_k. \tag{5.18}$$

We introduce the vector $\dot{g} = (\beta_1, \beta_2, \ldots)$. Clearly, $\dot{g} \in l_2$ since $\sum_{k=1}^{N} \beta_k^2 < \infty$. Relation (5.18) can therefore be rewritten in the form

$$2(\mathrm{Im}\,\dot{A})f = (f, \dot{g})\,\dot{g}. \tag{5.19}$$

Thus $\dot{g}$ is the channel element of A, which by virtue of (5.19) can be included in the operator complex

$$\dot{X} = (\dot{A},\ l_2,\ \dot{g},\ J = 1). \tag{5.20}$$

Let us now find the characteristic function of $\dot{A}$. To this end we consider the following equation for the value f_λ of the resolvent of $\dot{A}$ at the channel element $\dot{g}$:

$$(\dot{A} - \lambda I) f_\lambda = \dot{g} \tag{5.21}$$

or, in expanded form,

$$\begin{gathered}(\lambda_k - \lambda) f_\lambda (k) + i \sum_{j=k+1}^{N} f_\lambda (j) \beta_j \beta_k = \beta_k \\ (k = 1,\ 2, \ldots,\ N;\ N \leqslant \infty).\end{gathered} \tag{5.22}$$

Suppose there exists a solution of the system of equations (5.22) in the form

$$f_\lambda (k) = \frac{y (k) \beta_k}{\lambda_k - \lambda}. \tag{5.23}$$

Then $y (k)$ satisfies the equations

$$y (k) + i \sum_{j=k+1}^{N} y (j) \frac{\beta_j^2}{\lambda_j - \lambda} = 1, \qquad y (N) = 1. \tag{5.24}$$

Since $y (k - 1) - y (k) = -i \dfrac{\beta_k^2 y (k)}{\lambda_k - \lambda}$, we obtain the recursion formula

$$y (k - 1) = y (k) \frac{\bar{\lambda}_k - \lambda}{\lambda_k - \lambda}, \tag{5.25}$$

the repeated application of which implies

$$y(k-1) = y(N_0)\frac{\bar{\lambda}_{N_0}-\lambda}{\lambda_{N_0}-\lambda}\cdots\frac{\bar{\lambda}_k-\lambda}{\lambda_k-\lambda} \tag{5.26}$$

for $N_0 = k,\ k+1, \ldots, N$. In particular, when $N_0 = N$ $(N \leqslant \infty)$, we get

$$y(k-1) = \prod_{j=k}^{N} \frac{\lambda-\bar{\lambda}_j}{\lambda-\lambda_j} \quad (k = 1, 2, \ldots) \tag{5.27}$$

and hence

$$f_\lambda(k) = \frac{\beta_k}{\lambda_k-\lambda}\prod_{j=k+1}^{N}\frac{\lambda-\bar{\lambda}_j}{\lambda-\lambda_j}. \tag{5.28}$$

We now find

$$\dot{S}(\lambda) = 1 - i((\dot{A}-\lambda I)^{-1}\dot{g},\ \dot{g}) =$$
$$1 - i(f_\lambda,\ \dot{g}) = 1 - i\sum_{k=1}^{N}\frac{y(k)\beta_k^2}{\lambda_k-\lambda}. \tag{5.29}$$

From (5.24) it follows that $y(0) + i\sum_{k=1}^{N}\frac{y(j)\beta_j^2}{\lambda_j-\lambda} = 1$ and hence $\dot{S}(\lambda)$ has the multiplicative representation

$$\dot{S}(\lambda) = y(0) = \prod_{j=1}^{N}\frac{\bar{\lambda}_j-\lambda}{\lambda_j-\lambda}. \tag{5.30}$$

Thus the characteristic function of $\dot{A}$ coincides with the characteristic function (5.11) of A and hence, by Theorem 4.2, the principal components of the operator complexes $X = (A,\ H,\ g,\ J = 1)$ and $\dot{X} = (\dot{A},\ l_2,\ \dot{g},\ \ J = 1)$ are unitarily equivalent. It will be shown below that the complex $\dot{X}$ is simple.

The results obtained are summarized by the following:

Theorem 5.2. *Any simple operator complex with a one-dimensional imaginary component and* $l = 0$ *is unitarily equivalent to a triangular model of the form of* $\dot{X}$.

We note that the finite-dimensional subspaces H_n $(n = 1, 2, \ldots)$ of l_2 consisting of all elements satisfying the condition $f(k) = 0$ for $k > n$ are invariant under $\dot{A}$ and that the spectrum of $\dot{A}$ in H_n coincides with the numbers $\lambda_1, \lambda_2, \ldots, \lambda_n$. Consequently, the closed linear span of all of the finite-dimensional invariant subspaces coincides with l_2.

We require in the sequel an expression for the element $f = (\dot{A}^* - \lambda I)^{-1}\dot{g}$, which satisfies the equation

$$(\dot{A}^* - \lambda I) f = \dot{g} \tag{5.31}$$

or, in expanded form,

$$(\bar{\lambda}_k - \lambda) f(k) - i \sum_{j=1}^{k-1} f(j) \beta_j \beta_k = \beta_k \quad (k = 1, 2, \ldots). \tag{5.32}$$

Letting $y(k) = \dfrac{f(k)(\bar{\lambda}_k - \lambda)}{\beta_k}$, we obtain the equations

$$y(k) - i \sum_{j=1}^{k-1} y(j) \frac{\beta_j^2}{\bar{\lambda}_j - \lambda} = 1, \quad y(1) = 1. \tag{5.33}$$

Since

$$y(k) - y(k-1) = i \frac{\beta_{k-1}^2 y(k-1)}{\bar{\lambda}_{k-1} - \lambda},$$

we get

$$y(k) = \frac{\lambda_{k-1} - \lambda}{\bar{\lambda}_{k-1} - \lambda} y(k-1).$$

Consequently, $y(k) = \prod_{j=1}^{k-1} \dfrac{\lambda_j - \lambda}{\bar{\lambda}_j - \lambda}$ and hence

$$f(k) = [(\dot{A}^* - \lambda I)^{-1}\dot{g}]_k = \frac{\beta_k}{\bar{\lambda}_k - \lambda} \prod_{j=1}^{k-1} \frac{\lambda_j - \lambda}{\bar{\lambda}_j - \lambda}. \tag{5.34}$$

3. Let us now consider the case when the characteristic function has the form

$$S(\lambda) = e^{-i\int\limits_0^l \frac{ds}{\alpha(s)-\lambda}}, \tag{5.35}$$

i.e. when the spectrum of A is on the real axis.

In order to construct a triangular model we introduce the space $L_2(0,\ l)$ and define an operator $\dot{A}$ in the following manner:

$$(\dot{A}f)(x) = \alpha(x) f(x) + i\int\limits_x^l f(\xi)\, d\xi \quad (0 \leqslant x \leqslant l). \tag{5.36}$$

We will call this operator a *triangular model.*

Clearly, its adjoint is defined as follows:

$$(\dot{A}^* f)(x) = \alpha(x) f(x) - i\int\limits_0^x f(\xi)\, d\xi, \tag{5.37}$$

and hence

$$\left[\frac{1}{i}(\dot{A}-\dot{A}^*)f\right](x) = \int\limits_0^l f(\xi)\, d\xi = (f,\ \dot{g})\dot{g} \quad (\dot{g} \equiv 1,\ 0 \leqslant x \leqslant l). \tag{5.38}$$

As can be seen from (5.38), the operator $\dot{A}$ has a one-dimensional imaginary component and a channel element equal to the constant function one. It can be included in the model complex

$$\dot{X} = (\dot{A},\ L_2(0,\ l),\ \dot{g}(x) \equiv 1,\ J = 1). \tag{5.39}$$

Let us now calculate the element $f = (\dot{A} - \lambda I)^{-1}\dot{g}$, which satisfies the equation

$$([\dot{A} - \lambda I) f](x) \equiv 1 \quad (0 \leqslant x \leqslant l) \tag{5.40}$$

or, using (5.36),

$$[\alpha(x) - \lambda] f(x) + i \int_x^l f(\xi)\, d\xi \equiv 1. \tag{5.41}$$

We introduce the new unknown function $y(x) = [\alpha(x) - \lambda] f(x)$, which according to (5.41) satisfies the equation

$$y(x) + i \int_x^l \frac{y(\xi)}{\alpha(\xi) - \lambda}\, d\xi = 1 \quad (0 \leqslant x \leqslant l). \tag{5.42}$$

Clearly,

$$y(l) = 1. \tag{5.43}$$

Differentiating (5.42 with respect to x, we obtain the relation

$$y'(x) - i \frac{y(x)}{\alpha(x) - \lambda} = 0, \tag{5.44}$$

which together with (5.43) implies

$$y(x) = e^{-i \int_x^l \frac{d\xi}{\alpha(\xi) - \lambda}}.$$

But then

$$f(x) = [(\dot{A} - \lambda I)^{-1} \dot{g}](x) = \frac{1}{\alpha(x) - \lambda} e^{-i \int_x^l \frac{d\xi}{\alpha(\xi) - \lambda}} \quad (0 \leqslant x \leqslant l). \tag{5.45}$$

Since

$$y(0) = 1 - i \int_0^l \frac{y(s)\, ds}{\alpha(s) - \lambda} = 1 - i\, ((\dot{A} - \lambda I)^{-1} \dot{g}, \dot{g}) = \dot{S}(\lambda),$$

it follows that

$$\dot{S}(\lambda) = e^{-i\int_0^l \frac{ds}{\alpha(s)-\lambda}}. \tag{5.46}$$

Consequently, the characteristic functions of A and $\dot{A}$ coincide and hence, by Theorem 4.2, the principal components of $X = (A, H, g, J = 1)$ and $\dot{X} = (\dot{A}, L_2(0, l), \dot{g}(x) \equiv 1, J = 1)$ are unitarily equivalent.

The results obtained can be formulated in the form of a theorem.

Theorem 5.3. *If the characteristic function of A has the form* (5.35), *the principal component of the complex* $X = (A, H, g, J = 1)$ *is unitarily equivalent to the principal component of the model complex* $\dot{X} = (\dot{A}, L_2(0, l), \dot{g}, J = 1)$, *where A has the form* (5.36).

Let us now find an expression for the element $f = (\dot{A}^* - \lambda I)^{-1}\dot{g}$, which satisfies the equation

$$(\dot{A}^* - \lambda I) f \equiv 1 \quad (0 \leqslant x \leqslant l)$$

or, in more detail,

$$[\alpha(x) - \lambda] f(x) - i\int_0^x f(\xi)\, d\xi \equiv 1.$$

The function $y(x) = [\alpha(x) - \lambda] f(x)$ satisfies the equations

$$y(x) - i\int_0^x \frac{y(\xi)}{\alpha(\xi) - \lambda}\, d\xi = 1, \tag{5.47}$$

$$y(0) = 0. \tag{5.48}$$

Differentiating (5.47) with respect to x, we obtain the equation

$$y'(x) - \frac{i}{\alpha(x) - \lambda} y(x) = 0,$$

which together with (5.48) implies

$$y(x) = e^{i\int_0^x \frac{ds}{\alpha(s)-\lambda}}.$$

Therefore

$$f(x) = \frac{1}{\alpha(x) - \lambda} e^{i\int_0^x \frac{ds}{\alpha(s)-\lambda}}. \tag{5.49}$$

We now consider the special case when $\alpha(x) \equiv 0$, i.e. when the operator $\dot{A}$ is completely continuous. We have

$$\dot{A}f = i\int_x^l f(\xi)\,d\xi, \tag{5.50}$$

$$(\dot{A} - \lambda I)^{-1}\dot{g} = -\lambda^{-1}e^{i(l-x)\lambda^{-1}}, \tag{5.51}$$

$$S(\lambda) = e^{il\lambda^{-1}}, \tag{5.52}$$

$$(\dot{A}^* - \lambda I)^{-1}\dot{g} = -\lambda^{-1}e^{-ix\lambda^{-1}}, \tag{5.53}$$

In this case the complementary component is absent. In fact, the linear span of all elements of the form $\dot{A}^n\dot{g}$ $(n = 0, 1, 2, \ldots)$ contains all of the polynomials and hence is everywhere dense in $L_2(0, l)$.

4. In the general case, when the characteristic function $S(\lambda)$ has the form (5.10), we consider a triangular model obtained as a coupling of triangular models of forms (5.13) and (5.36) (and of the corresponding operator complexes of forms (5.20) and (5.39)).

The discrete part of the corresponding coupling of operator complexes has the form

$$\dot{X}_1 = \Big((\dot{A}_1 f_1)(k) = \Big(\alpha_k + \frac{i}{2}\beta_k^2\Big) f_1(k) + i\sum_{j=k+1}^{N} f_1(j)\,\beta_j\beta_k,\ \dot{H}_1 = l_2,$$
$$\dot{g}_1 = (\beta_1, \ldots, \beta_N),\ 1\Big), \tag{5.54}$$

while the continuous part has the form

$$\dot{X}_2 = \Big((\dot{A}_2 f_2)(x) = \alpha(x) f_2(x) + i\int_x^l f_2(\xi)\,d\xi,\ H_2 = L_2(0, l),$$
$$\dot{g}_2(x) \equiv 1,\ 1\Big). \tag{5.55}$$

Under a coupling of operator complexes the characteristic functions are multiplied in accordance with (4.9), and hence the characteristic function of the complex $\dot{X} = \dot{X}_2 \vee \dot{X}_1$ coincides with the given characteristic function $S(\lambda)$ of form (5.10). And this implies that the principal components of the complexes X and $\dot{X}$ are unitarily equivalent.

§2. OPERATORS OF ARBITRARY NON-HERMITIAN RANK

1. In Chapter IV we considered decreasing chains of invariant subspaces of the operator A. For the applications in the sequel it will be convenient to work also with increasing chains of invariant subspaces. We assume for the sake of simplicity that the non-Hermitian rank of A is finite.

Suppose $H' \subset H$ and $X' = (P'A, H', P'g_1, \ldots, P'g_r, J) = P'X$ is the projection of an operator complex (colligation) $X = (A, H, g_1, \ldots, g_r, J)$ onto H' (see Chapter I, §3). Let us agree to call the c.o.f. (or c.m.f.) of $P'X$ the *projection* onto H' of the characteristic function $S(\lambda)$ of X and to denote it by $P'[S(\lambda)]$.

Lemma 5.1. *Suppose* $H_0 = 0 \subset H_1 \subset H_2 \subset \cdots \subset H_n \subset \cdots$ *is an increasing chain of subspaces of the operator* A, *with* $\lim_{n\to\infty} P_n = I$ (P_n *is the orthogonal projection onto* H_n). *Then*

$$S(\lambda) = \lim_{n\to\infty} P_n[S(\lambda)], \tag{5.56}$$

where $S(\lambda)$ *is the c.o.f. of the complex* X. *The convergence in* (5.56) *is uniform in a neighborhood of each nonreal regular point of the function* $S(\lambda)$.

A proof of this lemma can be found in [5].

Theorem 5.4. *If in addition to the conditions of the preceding lemma all of the subspaces* H_n $(n = 1, 2, \ldots)$ *are invariant under* A, *the c.m.f.* $S(\lambda)$ *of* A *is representable in the form*

$$S(\lambda) = \overset{\curvearrowleft}{\prod_{k=1}^{\infty}} S_k(\lambda) = \lim_{n\to\infty}[S_n(\lambda)\cdots S_1(\lambda)], \tag{5.57}$$

where $S_k(\lambda) = P_k^{\perp}[S(\lambda)]$, *with* $P_k^{\perp}$ *denoting the orthogonal projection onto* $H_k \ominus H_{k-1}$.

Proof. Since $H_n \supset H_{n-1} \supset \cdots \supset H_1 \supset H_0 = 0$ is a decreasing chain of invariant subspaces and the c.m.f.'s are multiplied in reverse order to that of the corresponding operator functions, we get

$$P_n[S(\lambda)] = S_n(\lambda) \cdots S_1(\lambda). \tag{5.58}$$

Relation (5.57) follows from this result and Lemma 5.1.

2. We will cite without proof some well-known results for the multiplicative representations of c.m.f.'s and for the triangular representations of the corresponding operators, which generalize the results of §1 to the case of any non-Hermitian rank. The proofs can be found in [5].

Suppose we are given a scalar function $f(t)$ and a matrix function $B(t)$ on a segment $[a, b]$. We form a subdivision

$$a = t_0 < t_1 < \cdots < t_{n-1} < t_n = b,$$
$$t_k \leqslant \tau_{k+1} \leqslant t_{k+1} \ (k = 0, 1, 2, \ldots, n-1) \tag{5.59}$$

of $[a, b]$ and call the matrix

$$\overset{n}{\widehat{\prod_{k=1}}} e^{f(\tau_k)\Delta B(t_k)} = e^{f(\tau_n)\Delta B(t_n)} \cdots e^{f(\tau_2)\Delta B(t_2)} e^{f(\tau_1)\Delta B(t_1)}$$
$$(\Delta B(t_k) = B(t_k) - B(t_{k-1})) \tag{5.60}$$

an *integral product.* If the integral products tend to a definite limit as max $\Delta t_k \to 0$ independent of the choice of the subdivision of $[a, b]$ or the points t_k and τ_k, then this limit is called the *multiplicative Stieltjes integral* of $f(t)$ with weight $B(t)$ and is denoted by the symbol $\widehat{\int_a^b} e^{f(t)dB(t)}$.

If $f(t)$ is integrable in the Riemann sense and $B(t)$ satisfies the condition

$$\| B(t') - B(t'') \| \leqslant | t' - t'' | \quad (t', t'' \in [a, b])$$

then the integral $\widehat{\int_a^b} e^{f(t)dB(t)}$ exists.

Theorem 5.5. *The c.m.f.*

$$S(\lambda) = I - iJ \| ((A - \lambda I)^{-1} g_\alpha, g_\beta) \| \tag{5.61}$$

of an operator complex $X = (A, H, g_1, \ldots, g_r, J)$ *admits the representation*

$$S(\lambda) = \int_0^{l}\!\!\widehat{\ } e^{\frac{iJ}{\lambda - \alpha(t)} d\varepsilon(t)} \widehat{\prod_{k=1}^{\infty}} \left(I - \frac{iJ}{\lambda_k - \lambda} q_k^* q_k\right). \tag{5.62}$$

Let P *be the orthogonal projection onto the orthogonal complement of the span of all of the invariant subspaces of* A *corresponding to the nonreal points of its spectrum. Then*

$$l = \sum_{\alpha=1}^{r} \| P g_\alpha \|^2, \tag{5.63}$$

while $\alpha(t)$ *and* $\varepsilon(t)$ *are nondecreasing bounded functions on the segment* $[0, l]$, *the* λ_k $(k = 1, 2, \ldots)$ *are the nonreal points of the spectrum of* A *and the* q_k $(k = 1, 2, \ldots)$ *are row matrices satisfying the condition*

$$q_k J q_k^* = 2 \operatorname{Im} \lambda_k. \tag{5.64}$$

Here each of the numbers λ_k *is repeated as many times as the dimension of the corresponding invariant subspace, while the matrix function* $\varepsilon(t)$ *has the integral representation*

$$\varepsilon(t) = \int_0^t q^*(x) q(x)\, dx, \quad Sp q^*(x) q(x) \equiv 1, \tag{5.65}$$

where $q(x)$ *is a rectangular or square matrix whose rows are linearly independent at each point of some set of positive measure.*

3. Consider a Hilbert space $\dot{H} = \dot{H}_1 \oplus \dot{H}_2$, where $\dot{H}_1$ consists of sequences $f = \{f_1, f_2, \ldots\}$ of complex numbers satisfying the condition $\sum_{k=1}^{\infty} | f_k |^2 < \infty$, while the elements of $\dot{H}_2$ are vector functions

$f(x) = \| f_1(x), f_2(x), \ldots, f_p(x) \|, f_i(x) \in L_2(0, l) \ (i = 1, 2, \ldots, p)$

defined on a segment $[0, l]$.

We define the scalar product of elements $\tilde{f}^{(l)} = \{f_1^{(l)}, f_2^{(l)}, \ldots, f^l(x)\}$ $(l = 1, 2)$ of $\dot{H}$ by the formula

$$(\tilde{f}^{(1)}, \tilde{f}^{(2)}) = \sum_{k=1}^{\infty} \tilde{f}_k^{(1)} \overline{\tilde{f}_k^{(2)}} + \int_0^l f^{(1)}(x) \overline{f^{(2)}(x)}\, dx. \tag{5.66}$$

Let λ_k $(k = 1, 2, \ldots)$ denote a bounded sequence of nonreal numbers all of whose limit points lie on the real axis and let $\alpha(t)$ denote a bounded nondecreasing function on the segment $[0, l]$.

Let $q_k = \| q_k^{(1)}, q_k^{(2)}, \ldots, q_k^{(r)} \|$ $(k = 1, 2, \ldots, r \geqslant p)$ denote a sequence of row matrices with complex elements and let $q(x) = \| q_{ij}(x) \|$ denote a square or rectangular matrix of r columns and p rows that are linearly independent on a set of positive measure. It will be assumed in addition that the matrices q_k $(k = 1, 2, \ldots)$ and $q(x)$ have the properties

$$q_k J q_k^* = 2 \operatorname{Im} \lambda_k \ (k = 1, 2, \ldots), \tag{5.67}$$

$$\operatorname{Tr} q^*(x) q(x) = 1 \ (0 \leqslant x \leqslant l), \tag{5.68}$$

$$\sum_{k=1}^{\infty} q_k q_k^* < \infty. \tag{5.69}$$

We define an operator $\dot{A}\tilde{f}$ in H by putting

$$(\dot{A}\tilde{f})_k = \lambda_k f_k + i \sum_{j=k+1}^{\infty} f_j q_j J q_k^* + i \int_0^l f(t) q(t) J q_k^*\, dt, \tag{5.70}$$

$$(\dot{A}\tilde{f})(x) = \alpha(x) f(x) + i \int_x^l f(t) q(t) J q^*(x)\, dt \ (0 \leqslant x \leqslant l). \tag{5.71}$$

Any operator of the form (5.70)–(5.71) and satisfying the conditions enumerated above will be called a *triangular model.*

The operator $\dot{A}$ is defined on all of $\dot{H}$, is bounded and maps $\dot{H}_1$ into itself. If one introduces the elements

$$\dot{g}_\alpha = (\bar{q}_1^{(\alpha)}, \bar{q}_2^{(\alpha)}, \ldots) \quad (\dot{g}_\alpha \in \dot{H}_1, \ \alpha = 1, 2, \ldots r), \tag{5.72}$$

$$\dot{g}_\alpha(x) = (\bar{q}_{1\alpha}(x), \bar{q}_{2\alpha}(x), \ldots, \bar{q}_{p\alpha}(x)) \quad (\dot{g}_\alpha(x) \in \dot{H}_2) \tag{5.73}$$

and lets $\tilde{g}_\alpha = \dot{g}_\alpha + \dot{g}_\alpha(x)$ $(\tilde{g}_\alpha \in \dot{H})$, the operator $\dot{A}$ turns out to be included in a complex $\dot{X} = (\dot{A}, \tilde{H}, \tilde{g}_1, \ldots, \tilde{g}_r, J)$, which is an E-product $\dot{X} = \dot{X}_2 \vee \dot{X}_1$ of complexes $\dot{X}_i = (\dot{A}_i, \dot{H}_i, \dot{g}_{1i}, \ldots, \dot{g}_{ri}, J)$ $(i = 1, 2)$ in which the operators $\dot{A}_1$ and $\dot{A}_2$ are defined in $\dot{H}_1$ and $\dot{H}_2$ respectively by the equalities

$$(\dot{A}_1 f)_k = \lambda_k f_k + i \sum_{l=k+1}^{\infty} f_l q_l J q_k^*, \tag{5.74}$$

$$(\dot{A}_2 f)_k = \alpha(x) f(x) + i \int_x^l f(t) q(t) J q^*(x) \, dt, \tag{5.75}$$

while the channel elements are given by the equalities $\dot{g}_{\alpha 1} = \dot{g}_\alpha$, $\dot{g}_{\alpha 2} = \dot{g}_\alpha(x)$ $(\alpha = 1, 2, \ldots, r)$.

It can be verified by means of a simple calculation that the c.m.f.'s of $\dot{X}_1$ and $\dot{X}_2$ are respectively equal to

$$\dot{S}_1(\lambda) = \prod_{k=1}^{\overset{\frown}{\infty}} \left(I - \frac{iJ}{\lambda_k - \lambda} q_k^* q_k \right), \tag{5.76}$$

$$\dot{S}_2(\lambda) = \overset{\frown}{\int_0^l} e^{\frac{iJ}{\lambda - \alpha(t)} d\sigma(t)}. \tag{5.77}$$

By virtue of (5.62) the c.m.f. of $\dot{X}$ has the form

$$\dot{S}(\lambda) = \dot{S}_2(\lambda) \dot{S}_1(\lambda). \tag{5.78}$$

Let us agree to call the complex $\dot{X}$ a *triangular model* and to call the complexes $\dot{X}_1$ and $\dot{X}_2$ the *discrete* and *continuous* parts of $\dot{X}$ respectively. We note that the number of nonreal points in the spectrum of $\dot{A}$ can be finite, so that the case $\dim \dot{H}_1 < \infty$ is possible.

If $X = (A, H, g_1, \ldots, g_r, J)$ is an arbitrary complex, its c.m.f. has the form (5.62). With respect to the quantities q_k, $q(x)$, λ_k, $\alpha(t)$ and l in this representation we construct the triangular model $\dot{X} = (\dot{A}, \dot{H}, \tilde{g}_1, \ldots, \tilde{g}_r, J)$. Then Theorem 4.2 implies

Theorem 5.6. *The principal component of an arbitrary operator complex X is unitarily equivalent to the principal component of the corresponding triangular model* (5.70)–(5.71).

4. In working with a triangular model it is often necessary to determine the elements belonging to its complementary component. The following theorem contains some results in this direction [5].

Theorem 5.7. *The complementary component of a triangular model $\dot{X}$ coincides with the complementary component of its continuous part $\dot{X}_2$. If $\alpha(x) \equiv 0$ $(0 \leqslant x \leqslant l)$ and $J = I$, the complementary component of $\dot{X}_2$ consists of the elements $f(x)$ $(f(x) \in L_2(0, l))$ satisfying the condition*

$$f(x)q(x) \equiv 0 \quad (0 \leqslant x \leqslant l). \tag{5.79}$$

If, in particular, the rows of the matrix $q(x)$ are almost everywhere linearly independent, the complementary component reduces to zero.

5. Let $X = (A, H, g_1, \ldots, g_r, J)$ be a simple complex. We decompose the space H into the sum $H = H_1 \oplus H_2$ of two subspaces, the first of which coincides with the closed linear span of all of the invariant subspaces of A corresponding to the nonreal points of its spectrum. In conformity with this the c.m.f. is represented in the form (5.62). It is not difficult to see that the subspace H_1 coincides with H if and only if the quantity l in (5.62) is equal to zero.

In this case the continuous part of the model is absent. We can now easily obtain the following theorem [5].

Theorem 5.8. *If A is a dissipative completely non-self-adjoint operator,*

$$\sum_{k=1}^{\infty} \operatorname{Im} \lambda_k \leqslant \operatorname{Tr} \frac{A - A^*}{2i} \tag{5.80}$$

The system of finite-dimensional invariant subspaces of A is complete in H if and only if the equality sign holds in relation (5.80).

Remark. If $X = (A, H, g_1, \ldots, g_r, I)$ is an arbitrary dissipative complex and $\dot{X} = (\dot{A}, H, \tilde{g}_1, \ldots, \tilde{g}_r, I)$ is a triangular model of it, then by virtue of (4.65)

$$\operatorname{Tr} 2\,(\operatorname{Im} A) = \operatorname{Tr} 2\,(\operatorname{Im} \dot{A}) = \sum_{\alpha=1}^{r} \| \tilde{g}_\alpha \|^2 . \tag{5.81}$$

Using expressions (5.72) and (5.73) for the model channel elements and equality (5.81), we get

$$\operatorname{Tr} 2\,(\operatorname{Im} A) = \sum_{k=1}^{\infty} q_k q_k^* + \int_0^l \sum_{j=1}^{p} \sum_{\alpha=1}^{r} | q_{j\alpha}(x) |^2 \, dx =$$

$$\sum_{k=1}^{\infty} 2 \operatorname{Im} \lambda_k + \int_0^l \operatorname{Tr} q^*(x)\, q(x)\, dx.$$

This result together with equality (5.68) implies

$$\operatorname{Tr} 2\,(\operatorname{Im} A) = \sum_{k=1}^{\infty} 2 \operatorname{Im} \lambda_k + l . \tag{5.82}$$

Conditions for the completeness of a system of generalized eigenvectors have been studied for dissipative operators in [15],[20],[29],[31],[33]. We cite one of the more recent results obtained in this area [20].

Theorem 5.9. *Let A be a dissipative operator in H, let* $\{\lambda_1, \lambda_2, \ldots, \lambda_n, \ldots\}$ *be a sequence of eigenvalues of A, let* m_n *be the index of the eigenvalue* λ_n *and let* H_n *be the corresponding generalized eigenspace. If the condition*

$$\prod_{\substack{k=1\\(k\neq j)}}^{\infty}\left|\frac{\lambda_j-\lambda_k}{\lambda_j-\bar{\lambda}_k}\right|^{m_j m_k}>\delta\quad(j=1,2,3,\ldots)\tag{5.83}$$

is satisfied for some $\delta>0$, *then any system* $\{g_1^{(n)},\ldots,g_{m_n}^{(n)}\}$, *where each* $\{g_1^{(n)},\ldots,g_{m_n}^{(n)}\}$, *is an orthonormal basis for* H_n, *forms a basis equivalent to an orthonormal basis.*

We note that condition (5.83) is equivalent to the conditions

$$\sup_{1<j<\infty}\sum_{\substack{k=1\\(k\neq j)}}^{\infty} m_j m_k\,\frac{\operatorname{Im}\lambda_j\cdot\operatorname{Im}\lambda_k}{|\lambda_j-\bar{\lambda}_k|^2}<\infty,\tag{5.84}$$

$$\sup_{\substack{1<j,\,k<\infty\\ j\neq k}}\frac{4\operatorname{Im}\lambda_j\cdot\operatorname{Im}\lambda_k}{|\lambda_j-\bar{\lambda}_k|^2}<1.\tag{5.85}$$

§3. COMPLETE DISSIPATIVE OPERATORS AND COMPLEXES

A dissipative operator A defined in H is said to be *complete* if the closed linear span of the invariant subspaces corresponding to the nonreal points of its spectrum coincides with H.

From the results presented in §2 it follows that a completely non-self-adjoint dissipative operator with finite non-Hermitian rank (or with a finite value of the trace $\operatorname{Tr}(\operatorname{Im}A)$) will be complete if and only if its c.o.f. is representable in the form

$$S(\lambda)=\widehat{\prod_{k=1}^{\infty}}\left(I-\frac{i}{\lambda-\lambda_k}q_k^*q_k\right).\tag{5.86}$$

By virtue of formula (4.77) this condition is equivalent to the condition

$$\det S(\lambda)=\prod_{k=1}^{\infty}\frac{\lambda-\bar{\lambda}_k}{\lambda-\lambda_k}.\tag{5.87}$$

Lemma 5.2. *If the complex* $X = (A, H, g_1, \ldots, g_r, J)$ *is an E-product of the complexes* $X_2 = (A_2, H_2, g_{12}, \ldots, g_{r2}, J)$ *and* $X_1 = (A_1, H_1, g_{11}, \ldots g_{r1}, J)$ *and* A_1 *is a complete dissipative operator, the complementary component of* X *lies in* H_2.

Proof. Since A is a complete dissipative operator, there exists an orthonormal basis e_k in H_1 such that

$$\begin{aligned} A_1 e_1 &= \lambda_1 e_1, \\ A_1 e_2 &= a_{21} e_1 + \lambda_2 e_2, \\ A_1 e_3 &= a_{31} e_1 + a_{32} e_2 + \lambda_3 e_3, \\ &\ldots\ldots\ldots\ldots \end{aligned} \tag{5.88}$$

Let P_0 be the orthogonal projection onto the complementary component H_0. Then $AP_0 = P_0 A$ and

$$AP_0 e_1 = P_0 A e_1 = P_0 A_1 e_1 = \lambda_1 P_0 e_1.$$

Thus $P_0 e_1$ is an eigenvector of A corresponding to the nonreal eigenvalue λ_1. Since $A = A^*$ on H_0, it follows that $P_0 e_1 = 0$. Analogously,

$$AP_0 e_2 = P_0 A e_2 = P_0 A_1 e_2 = \lambda_2 P_0 e_2$$

and hence $P_0 e_2 = 0$. Continuing in this way, we arrive at the conclusion that the complementary component is orthogonal to H_1.

Corollary. *If* A *is a complete dissipative operator, its complementary component is equal to zero, i.e. a complete dissipative operator is always completely non-self-adjoint.*

Let us agree to call a complex $X = (A, H, g_1, \ldots, g_r, J)$ *complete* if A is a complete operator.

Lemma 5.3. *If* $X = (A, H, g_1, \ldots, g_r, I)$ *is a complete dissipative complex, its adjoint* $X^* = (-A^*, H, g_1, \ldots, g_r, I)$ *is also complete and dissipative.*

Proof. From formula (4.23) it follows that the relation

$$S_{X^*}(\lambda) = S_X^*(-\bar{\lambda}) \tag{5.89}$$

exists between the c.m.f.'s of X and X^*. Since the eigenvalues of an operator are the poles of its c.o.f., this relation implies that the nonreal spectrum of the operator $-A^*$ consists of the points $-\bar{\lambda}_1, -\bar{\lambda}_2, \ldots$ if the nonreal spectrum of A consists of $\lambda_1, \lambda_2, \ldots$. The points $-\bar{\lambda}_1, -\bar{\lambda}_2, \ldots$ clearly satisfy the completeness condition in Theorem 5.8.

Theorem 5.10. *An E-product $X = X_2 \vee X_1$ of two complete dissipative complexes is a complete dissipative complex.*

Proof. Let H_0 be the complementary component of X. By Lemma 5.2 we have $H_0 \subset H_2$. Since $X^* = X_1^* \vee X_2^*$ it follows by virtue of Lemmas 5.2 and 5.3 that $H_0 \subset H_1$. Therefore $H_0 = 0$ and hence X is a simple complex. Since to a product of complexes there corresponds the product of their c.o.f.'s we get, going over to determinants,

$$\det S(\lambda) = \det S_1(\lambda) \cdot \det S_2(\lambda). \tag{5.90}$$

But this result implies that if the completeness condition (5.87) holds for X_1 and X_2 it also holds for X.

§4. UNIVERSAL MODELS OF COMPLETE DISSIPATIVE COMPLEXES

Let $\Lambda = (\lambda_1, \lambda_2, \ldots)$ be a given λ-set. A complete dissipative complex $X = (A, H, g_1, \ldots, g_r, I)$ will be said to belong to the class $D_r(\Lambda)$ if the nonreal spectrum of its operator is contained in Λ. From the results of Chapter IV, §3 it follows that any complete dissipative complex X with r channel elements belongs to the class $D_r(\Lambda(A))$, where $\Lambda(A)$ is the nonreal spectrum of its operator A.

With respect to the given set X we construct in the same way as was done in §1 a triangular model $\dot{X}_1 = (\dot{A}_1, l_2^{(1)}, \dot{g}_1, 1)$ of form (5.20), after which we form the orthogonal sums

$$\dot{A} = \dot{A}_1 \oplus \dot{A}_2 \oplus \cdots \oplus \dot{A}_r, \tag{5.91}$$

$$\dot{H} = l_2^{(1)} \oplus l_2^{(2)} \oplus \cdots \oplus l_2^{(r)}, \tag{5.92}$$

in which the $\dot{A}_\alpha$ and $l_2^{(\alpha)}, (\alpha = 1, 2, \ldots, r)$ are r copies of the operator $\dot{A}_1$ and the space l_2^1 respectively.

Thus the elements f of $\dot{H}$ have the form $f = (f_1, \ldots, f_r)$ $(f_\alpha \in l_2)$, so that $\dot{A}f = (\dot{A}_1 f_1, \ldots, \dot{A}_1 f_r)$. We will call the complex

$$\Theta_r(\Lambda) = (\dot{A}, \dot{H}, \dot{g}_1, \ldots, \dot{g}_r, I), \tag{5.93}$$

where

$$\dot{g}_1 = (\dot{g}_1, 0, \ldots, 0), \ \dot{g}_2 = (0, \dot{g}_1, \ldots, 0), \ \ldots, \ \dot{g}_r = (0, 0, \ldots, g_1),$$

a *universal complex of class* D_r. Clearly, $\Theta_r(\Lambda) \in D_r(\Lambda(A))$. From the expression (5.30) for the characteristic function of a triangular model it follows that the c.m.f. $S_\Theta(\lambda)$ of $\Theta_r(\Lambda)$ has the form

$$S_\Theta(\lambda) = \prod_{k=1}^{\infty} \frac{\lambda - \bar{\lambda}_k}{\lambda - \lambda_k} I. \tag{5.94}$$

Theorem 5.11. *All of the complexes of class $D_r(\Lambda)$ can be obtained up to unitary equivalence by projecting the complex $\Theta_r(\Lambda)$ onto the invariant subspaces of the operator $\dot{A}$.*

We preface the proof of Theorem 5.11 with a lemma.

Lemma 5.4. *Suppose the c.m.f. $S_X(\lambda)$ of a simple complex $X = (A, H, g_1, \ldots, g_r, J)$ has been represented in the form of a product*

$$S_X(\lambda) = S_2(\lambda) S_1(\lambda), \tag{5.95}$$

in which $S_1(\lambda)$ and $S_2(\lambda)$ are the c.m.f.'s of simple complexes X_1 and X_2 respectively. If the product $X' = X_2 \vee X_1$ of these complexes is a simple complex, the c.m.f. $S_1(\lambda)$ is the projection of $S_X(\lambda)$ onto an invariant subspace of the operator A.

Proof. Formula (4.9) implies $S_{X'}(\lambda) = S_2(\lambda) S_1(\lambda) = S_X(\lambda)$ and hence the complexes X and X' are unitarily equivalent. On the other hand, from the definition of an E-product we have $X_1 = P_1' X'$, where P_1' in the orthogonal projection of H' onto an invariant subspace H_1', and hence $S_1(\lambda) = P_1'[S_{X'}(\lambda)]$. Therefore, under the unitary mapping $H = UH'$ we obtain the space $H_1 = UH' = UP_1'H' = UP_1'U^{-1}H = P_1 H$, so that $S_1(\lambda) = P_1[S_X(\lambda)]$.

Proof of Theorem 5.11. Suppose $X = (A, H, g_1, \ldots, g_r, I) \in D_r(\Lambda)$. According to (5.86) we have

$$S_X(\lambda) = \overset{\infty}{\overset{\frown}{\prod_{k=1}}} \left(I - \frac{i}{\lambda - \mu_k} q_k^* q_k\right) \quad (\mu_k \in \Lambda). \tag{5.96}$$

It follows from equality (4.85) that

$$S_\Theta(\lambda) S_X^{-1}(\lambda) = \overset{\infty}{\overset{\frown}{\prod_{k=1}}} \left(\left(I - \frac{i}{\lambda - \mu_k} q_k^* q_k\right)^{-1} \frac{\lambda - \bar{\mu}_k}{\lambda - \mu_k}\right) = \overset{\infty}{\overset{\frown}{\prod_{k=1}}} S_k^\perp(\lambda), \tag{5.97}$$

where $S_k^\perp(\lambda) \in \Omega_I$ is the c.m.f. of some dissipative complex. Since this product converges uniformly in a neighborhood of each regular point of the function $S_X(\lambda)$, we get $\overset{\infty}{\overset{\frown}{\prod_{k=1}}} S_k^\perp(\lambda) \in \Omega_I$.

Thus

$$S_\Theta(\lambda) = S_2(\lambda) S_X(\lambda), \tag{5.98}$$

where

$$S_2(\lambda) = \overset{\infty}{\overset{\frown}{\prod_{k=1}}} S_k^\perp(\lambda) \in \Omega_I .$$

Since $S_2(\lambda) \in \Omega_I$ it follows from Theorem 4.3 that $S_2(\lambda)$ is the c.m.f. of some simple dissipative complex X_2. The completeness condition (5.87) implies

$$\det S_2(\lambda) = \frac{\det S_\Theta(\lambda)}{\det S_X(\lambda)} = \prod_{k=1}^{\infty} \left(\frac{\lambda - \bar{\mu}_k}{\lambda - \mu_k}\right)^{r-1} \tag{5.99}$$

and therefore $X_2 \in D_r(\Lambda)$. By virtue of Theorem 5.10 and Lemma 5.4 we have $S_X(\lambda) = P_1[S_\Theta(\lambda)]$, where P_1 is the orthogonal projection onto an invariant subspace H_1 of the operator $\dot{A}$. Letting X_1 denote the projection $P_1\Theta_r(\Lambda)$ of $\Theta_r(\Lambda)$ onto H_1, we get $S_{X_1}(\lambda) = P_1[S_\Theta(\Lambda)] = S_X(\lambda)$, which implies the unitary equivalence of the complexes X and X_1. The theorem is proved.

§5. A UNIVERSAL MODEL WITH SPECTRUM AT THE ORIGIN

If the entire spectrum of the operator A included in a complex $X = (A, H, g_1, \ldots, g_r, J)$ reduces to the point $\lambda = 0$, the c.m.f. has by virtue of (5.62) the form

$$S_x(\lambda) = \overset{l}{\underset{0}{\int^{\frown}}} e^{i\lambda^{-1}d_\varepsilon(t)}. \tag{5.100}$$

Since $\| \varepsilon(t_2) - \varepsilon(t_1) \| \leqslant | t_2 - t_1 |$ by virtue of (5.65),

$$\| S_x(\lambda) \| \leqslant e^{|\lambda|^{-1}} \int_0^l dt = e^{|\lambda|^{-1}l}. \tag{5.101}$$

Relation (5.82) implies in the present case that

$$l = 2\,\mathrm{Tr}\,(\mathrm{Im}\, A) \tag{5.102}$$

and hence that $S_X(\lambda)$ is an entire function of λ^{-1} of exponential type not greater than $l = 2\,\mathrm{Tr}\,(\mathrm{Im}\, A)$.

Let us agree to say that a simple complex $X = (A, H, g_1, \ldots, g_r, J)$ *belongs to the class* $C_r(a)$ $(0 < a < \infty)$ if $2\,\mathrm{Tr}\,(\mathrm{Im}\, A) \leqslant a$ and the spectrum of A consists of only the point $\lambda = 0$.

We consider a complex of form (5.55) when $\alpha(x) \equiv 0$ $(0 \leqslant x \leqslant l)$:

$$\Theta_1(l) = (\dot{B}_1 f = i \int_x^l f(t)\,dt,\; L_2^{(1)}(0, l),\; \dot{g}_1(x) \equiv 1 \;(0 \leqslant x \leqslant l),\; J = 1) \tag{5.103}$$

and form the orthogonal sums

$$\dot{B} = \dot{B}_1 \oplus \dot{B}_2 \oplus \cdots \oplus \dot{B}_r;$$
$$L_2 = L_2^{(1)}(0, l) \oplus L_2^{(2)}(0, l) \oplus \cdots \oplus L_2^{(r)}(0, l),$$

where the $\dot{B}_\alpha$ and $L_2^{(\alpha)}(0, l)$ ($\alpha = 1, 2, \ldots, r$) are r copies each of the operator $\dot{B}_1$ and the space $L_2(0, l)$ respectively. Thus the elements $\overrightarrow{f(x)}$ of $\dot{L}_2$ have the form $f(\vec{x}) = (f_1(x), \ldots, f_r(x))$ $(f_\alpha(x) \in L_2(0, l))$, while the operator B has the form

$$\dot{B}f = i\left(\int_x^l f_1(t)\,dt, \ldots, \int_x^l f_r(t)\,dt\right) \quad (0 \leqslant x \leqslant l). \tag{5.104}$$

The complex

$$\Theta_r(l) = (\dot{B}, L_2, \dot{g}_1(x), \ldots, \dot{g}_r(x), I), \tag{5.105}$$

where

$$\dot{g}_1(x) \equiv (1, 0, \ldots, 0), \ldots, g_r(x) \equiv (0, 0, \ldots, 1) \quad (0 \leqslant x \leqslant l),$$

will be called a *universal complex of class* C_r.

Clearly, $\Theta_r(l) \in C_r(rl)$. The c.m.f. $S_\Theta(\lambda)$ of $\Theta_r(l)$ has the form

$$S_\Theta(\lambda) = e^{il\lambda^{-1}}I. \tag{5.106}$$

Theorem 5.12. *All of the complexes of class* $C_r(l)$ *can be obtained up to unitary equivalence by projecting the complex* $\Theta_r(l)$ *onto the invariant subspaces of the operator* $\dot{B}$.

A proof of this theorem (for a wider class $\Lambda^{(\exp)}$) can be found in [4].

It follows in particular from Theorem 5.12 that the internal operator A of a complex belonging to the class $C_r(l)$ is completely continuous.

Remark. Consider the class $K_r(\Lambda, l)$ of all complexes of the form $X = X_2 \vee X_1$, where $X_1 \in D_r(\Lambda)$ and $X_2 \in C_r(l)$. The complex $\Theta_r(\Lambda, l) = \Theta_r(l) \vee \Theta_r(\Lambda)$ is a *universal complex of class* K_r in the sense that a theorem analogous to Theorems 5.11 and 5.12 is valid for it: *all of the complexes of class* $K_r(\Lambda, l)$ *can be obtained up to unitary equivalence by projecting the complex* $\Theta_r(\Lambda, l)$ *onto the invariant subspaces of its internal operator.* The proof is analogous to the proofs of Theorems 5.11 and 5.12.

§6. THE ASYMPTOTIC STABILITY OF SOME CLASSES OF DIFFERENTIAL EQUATIONS IN HILBERT SPACE

Consider the abstract Cauchy problem for a differential equation in Hilbert space

$$i\frac{dh}{dt} + Ah = 0,$$
$$h(0) = h_0, \tag{5.107}$$

where A is a bounded dissipative operator.

The relation

$$\frac{d\|h\|^2}{dt} = \left(\frac{dh}{dt}, h\right) + \left(h, \frac{dh}{dt}\right) = -2((\mathrm{Im}\,A)h, h) \leqslant 0 \tag{5.108}$$

implies the existence of the limit $\lim\limits_{t\to\infty} \|h(t)\|$.

If $\lim\limits_{t\to+\infty} \|h(t)\| = 0$ for the solution of problem (5.107) under any initial value h_0, problem (5.107) is said *to be asymptotically stable.*

The general solution of problem (5.107) has the form

$$h(t) = e^{iAt}h_0. \tag{5.109}$$

We will require the following theorem in the sequel.

Theorem 5.13. *If A is a completely nonselfadjoint and completely continuous dissipative operator, problem* (5.107) *is asymptotically stable:*

$$\lim_{t\to+\infty} \|e^{iAt}h_0\| = 0 \quad (h_0 \in H). \tag{5.110}$$

A proof of this theorem can be found in [33].

One can easily prove the following

Theorem 5.14. *If A is a complete dissipative operator, problem* (5.107) *is asymptotically stable.*

Proof. Since A is a complete operator, there exists a sequence of finite-dimensional invariant subspaces $H^{(n)}$ such that $\lim_{n\to\infty} P^{(n)} = I$, where $P^{(n)}$ is the orthogonal projection onto $H^{(n)}$. If the initial value $h_0' \in H^{(n)}$, the solution of problem (5.107) has the form $h'(t) = \sum_{k=1}^{m} Q_k(t) e^{i\lambda_k t}$, where the $Q_k(t)$ are polynomials and $\operatorname{Im}\lambda_k > 0$. It then follows that $\lim_{t\to+\infty} \|h'(t)\| = 0$. Suppose now h_0 is an arbitrary element of H. There exists an n and an $h_0' \in H^{(n)}$ such that $\|h^0 - h_0'\| < \varepsilon$. But it then follows by virtue of (5.108) that $\|h(t)\| \leqslant \|h'(t)\| + \|h(t) - h'(t)\| \leqslant \|h'(t)\| + \|h(0) - h'(0)\| \leqslant \|h'(t)\| + \varepsilon.$ Choosing N_ε so that $\|h'(t)\| < \varepsilon \quad (t > N_\varepsilon)$, we get $\|h(t)\| < 2\varepsilon \quad (t > N_\varepsilon)$ Q.E.D.

Theorem 5.15. *If the complex* $X = (A, H, g_1, \ldots, g_r, J)$ *belongs to the class* $K_r(\Lambda, l)$, *problem* (5.107) *is asymptotically stable.*

The proof of Theorem 5.15 is left to the reader.

CHAPTER VI

STOCHASTIC FIELDS AND STOCHASTIC PROCESSES

§1. BASIC CONCEPTS[1]

Consider a set Ω of elements (*elementary events*) and a σ-algebra of subsets of Ω. Let $\mu(\mathfrak{M})$ be a countably additive measure that is defined on the σ-algebra and satisfies the conditions

$$1)\ \mu(\mathfrak{M}) \geqslant 0,$$
$$2)\ \mu(\Omega) = 1. \tag{6.1}$$

Such a normalized measure is called a *probability measure*.

A *stochastic function (random variable)* is any measurable (relative to the σ-algebra) complex valued function $z(\omega)$ defined on the set of elementary events $\omega \in \Omega$. If a random variable depends not only on $\omega \in \Omega$ but also on a

[1] In §§1-3 we present information from the theory of stochastic processes that is needed for the sequel. We have made use here of the monographs of Rozanov [36] and Itô [18]. The results of §§4-5 are due to M. S. Livshits and A. A. Yantsevich.

point $x \in R_n$, it is called a *stochastic field.*[1] In the particular case $n = 1$ we arrive at the notion of a stochastic process $z(t, \omega)$, $t \in R_1$.[2]

The most important characteristics of a stochastic process are the *mathematical expectation* $Mz(t)$ and the *correlation function* $v(t, s) = Mz(t)\,z(s)$, which are defined as follows:

$$Mz(t) = \int_{\Omega} z(t, \omega)\,\mu(d\omega); \tag{6.2}$$

$$v(t, s) = \int_{\Omega} z(t, \omega)\,\overline{z(s, \omega)}\,\mu(d\omega). \tag{6.3}$$

But the calculation of either the mathematical expectation or the correlation function by formulas (6.2) or (6.3) requires the construction of the probability measure $\mu(\mathfrak{M})$, which in most problems is not known. It turns out that a procedure for finding the mathematical expectation can be realized by making use of the so-called n-variate distribution functions.

By a *distribution function* $F_t(\xi)$ of a real-valued stochastic process is meant the function $\mu(z(t) < \xi)$. By an *n-variate distribution function* $F_{t_1 t_2 \ldots t_n}(\xi_1, \xi_2, \ldots, \xi_n)$ is meant the function

$$\mu(z(t_1) < \xi_1,\ z(t_2) < \xi_2, \ldots, z(t_n) < \xi_n), \tag{6.4}$$

where $t_1, t_2, \ldots, t_n$ are any elements of R_1.

It is obvious that the distribution functions must satisfy the following conditions:

1) the symmetry condition

$$F_{t_{i_1} \ldots t_{i_n}}(\xi_{i_1}, \ldots, \xi_{i_n}) = F_{t_1, \ldots, t_n}(\xi_1, \ldots, \xi_n), \tag{6.5}$$

where $i_1, \ldots, i_n$ is any permutation of the numbers $1, 2, \ldots, n$, and

[1] Translators note: In addition, one must make some continuity or measurability assumption about the nature of $z(t, \omega)$ as a function of t.

[2] The dependence on ω will not be denoted in this case.

2) the consistency condition

$$F_{t_1 t_2 \ldots t_m t_{m+1} \ldots t_n}(\xi_1, \xi_2, \ldots, \xi_m, \infty, \ldots, \infty) = F_{t_1 \ldots t_m}(\xi_1, \ldots, \xi_m). \tag{6.6}$$

A.N. Kolmogorov showed [23] that a system of distribution functions (6.4) satisfying conditions (6.5) and (6.6) for any n uniquely determines a stochastic process. The proof of this fundamental fact is based on the possibility of extending a family of measures in a finite-dimensional space to a measure in an infinite-dimensional space.

From (6.2)–(6.4) it follows that

$$Mz(t) = \int_{-\infty}^{\infty} \xi \, dF_t(\xi), \tag{6.7}$$

$$v(t, s) = \iint_{-\infty}^{\infty} \xi_1 \xi_2 \, d^2 F_{ts}(\xi_1, \xi_2). \tag{6.8}$$

Thus we can determine $Mz(t)$ and $v(t, s)$ if we know $F_t(\xi)$ and $F_{ts}(\xi_1, \xi_2)$. But the construction of $F_t(\xi)$ and $F_{ts}(\xi_1, \xi_2)$ is much simpler than a direct determination of the probability measure.

The study of random variables and stochastic processes is facilitated by making use of the methods of Hilbert space theory. Consider a probability space Ω and the random variables $z(\omega)$ on Ω satisfying the condition

$$M|z(\omega)|^2 < \infty. \tag{6.9}$$

Let $L_2(\Omega)$ denote the Hilbert space of such $z(\omega)$ with the scalar product

$$(z_1, z_2) = Mz_1(\omega)\overline{z_2(\omega)} = \int_{\Omega} z_1(\omega)\overline{z_2(\omega)}\,\mu(d\omega). \tag{6.10}$$

If $z(t)$ is a stochastic process and $M|z(t)|^2 < \infty$, the closed linear span of the values of $z(t)$ for $t \in R_1$ is a Hilbert space $H \subset L_2(\Omega)$. Clearly, the process $z(t)$ can be regarded as a curve in either H_z or $L_2(\Omega)$, while the correlation function $v(t, s)$ of $z(t)$ is a scalar product:

$$v(t, s) = (z(t), z(s)). \tag{6.11}$$

We note that the correlation function is a *positive semidefinite self-adjoint kernel*, i.e.

$$M\left|\sum_{k=1}^{n} z(t_k)\xi_k\right|^2 = \sum_{j,k=1}^{n} v(t_k, t_j)\xi_k\bar{\xi}_j \geqslant 0 \tag{6.12}$$

$$v(t, s) = \overline{v(s, t)} \tag{6.13}$$

Lemma 6.1. *Suppose the correlation functions of a given pair of processes $z_1(t)$ and $z_2(t)$ coincide, i.e.*

$$v_{z_1}(t, s) = v_{z_2}(t, s) \tag{6.14}$$

and suppose $z_1(t)$ and $z_2(t)$ are the corresponding curves in the Hilbert spaces $H_{z_1} \subset L_2(\Omega)$ and $H_{z_2} \subset L_2(\Omega)$ spanned by the values of $z_1(t)$ and $z_2(t)$ respectively. Then $z_1(t) = Uz_2(t)$, where U is an isometric operator acting in $L_2(\Omega)$ and mapping H_{z_2} onto H_{z_1}.

Proof. We first define the operator U on the linear combinations $\sum_k c_k z_2(t_k)$:

$$U\left(\sum_k c_k z_2(t_k)\right) = \sum_k c_k z_1(t_k). \tag{6.15}$$

This operator is isometric by virtue of (6.14), it can be extended by continuity onto all of H_{z_2} and its range coincides with all of H_{z_1}. But then (6.15) implies that $Uz_2(t) = z_1(t)$.

§2. HOMOGENEOUS STOCHASTIC FIELDS AND STATIONARY STOCHASTIC PROCESSES

We will now consider some special classes of stochastic fields and processes.

A stochastic field $z(x)$, $x \in R_n$, is said to be *homogeneous* if its correlation function $v(x, y) = Mz(x)\overline{z(y)}$ depends only on the difference of the arguments:

$$Mz(x)\overline{z(y)} = v(x - y).$$

For example, a field of the form

$$z(x) = e^{i\sum_{k=1}^{n} A_k x^k} z_0, \tag{6.16}$$

where z_0 is a fixed element of the Hilbert space H_z spanned by the values of $z(x)$ and the A_k $(k = 1, 2, \ldots, n)$ are commuting Hermitian operators in H_z, is homogeneous. In fact, calculating $v(x, y)$, we get

$$v(x, y) = (z(x), z(y)) = \left(e^{i\sum_{k=1}^{n} A_k x^k} z_0, e^{i\sum_{k=1}^{n} A_k y^k} z_0\right) =$$
$$\left(e^{-i\sum_{k=1}^{n} A_k^* y^k} e^{i\sum_{k=1}^{n} A_k x^k} z_0, z_0\right) = \left(e^{i\sum_{k=1}^{n} A_k (x^k - y^k)} z_0, z_0\right) = v(x - y).$$

Definition. A stochastic field $z(x)$ is said to be *linearly representable* if in the Hilbert space H_z spanned by its values it can be represented in the form

$$z(x) = e^{i\sum_{k=1}^{n} A_k x^k} z_0, \tag{6.17}$$

where the A_k $(k = 1, 2, \ldots, n)$ are commutating not necessarily self-adjoint operators in H_z and z_0 is a fixed element of H_z.

Definition. By the *spectrum of a linearly representable process* will be meant the spectrum of the operator A appearing in the representation

$$z(t) = e^{itA} z_0 \,. \tag{6.18}$$

We proceed to the notion of a stationary process. A stochastic process is said to be *stationary in the broad sense* if its correlation function satisfies the condition

$$v(t, s) = v(t - s).$$

It will be assumed below that the correlation function of a stationary process is continuous.

Examples. 1. It is easily verified that the process

$$z(t) = \sum_{k=1}^{n} c_k e^{i\lambda_k t}, \tag{6.19}$$

where the c_k are independent random variables and the λ_k are arbitrary real numbers, is stationary.

2. A process of the form

$$z(t) = e^{itA} z_0, \tag{6.20}$$

where A is a self-adjoint operator in the Hilbert space H_z spanned by the values of $z(t)$ and z_0 is a fixed element of H_z, is obviously stationary.

Lemma 6.2. *If the correlation functions of a given pair of stochastic processes $z_1(t)$ and $z_2(t)$ coincide and one of these processes is linearly representable in the Hilbert space spanned by its values then so is the other.*

The proof follows directly from the lemma of §1. In fact, $Uz_2(t) = z_1(t)$, since $v_{z_1}(t, s) = v_{z_2}(t, s)$. Suppose now that the process $z_1(t)$ is linearly representable, i.e. $z_1(t) = e^{itA} z_1(0)$. Then $Uz_2(t) = e^{itA} z_1(0)$. Hence $z_2(t) = U^{-1} e^{itA} z_1(0)$. Putting $U^{-1} z_1(0) = z_{20}$, $\widetilde{A} = U^{-1} A U$ in this relation, we get $z_2(t) = e^{it\widetilde{A}} z_{20}$, i.e. $z_2(t)$ is linearly representable.

We define a family of operators U_τ in the Hilbert space H_z spanned by the values of a given stationary process $z(t)$ $(-\infty < t < \infty)$ as follows:

$$\begin{aligned} &U_\tau z(t) = z(t + \tau); \\ &U_\tau \sum_k c_k z(t_k) = \sum_k c_k z(t_k + \tau). \end{aligned} \tag{6.21}$$

These relations determine U_τ as a linear and isometric operator on the linear span of the curve $z(t)$. If we now take the closure of this linear span, we can extend the operator U_τ with preservation of linearity and isometry onto all of H_z, in which case, clearly, $U_\tau H_z = H_z$. Consequently, U_τ is a unitary operator in H_z.

The operators U_τ $(-\infty < \tau < \infty)$ form a group, since

$$U_{t+s} = U_t U_s = U_s U_t, \quad U_{-t} = U_t^{-1}. \tag{6.22}$$

But then, according to a theorem of Stone [1], one has the representation

$$U_t = e^{itA}, \tag{6.23}$$

where A is a generally unbounded self-adjoint operator in H_z.

Since $z(t) = U_t z(0)$,

$$z(t) = e^{itA} z(0). \tag{6.24}$$

If we now make use of the spectral resolution of a self-adjoint operator:

$$A = \int_{-\infty}^{\infty} \lambda \, dE_\lambda, \tag{6.25}$$

where E_λ is a family of orthogonal projections in H_z, we can obtain for $z(t)$ the representation

$$z(t) = \int_{-\infty}^{\infty} e^{i\lambda t} \Phi(d\lambda), \tag{6.26}$$

in which $\Phi(d\lambda) = dE_\lambda z(0)$ is a so-called stochastic spectral measure, i.e. a stochastic set function on the real line satisfying the relation

$$(\Phi(\Delta_1), \Phi(\Delta_2)) = M\Phi(\Delta_1)\overline{\Phi(\Delta_2)} = 0$$

if $\Delta_1 \cap \Delta_2 = 0$.

The representation (6.26) of a stationary stochastic process is called the *spectral resolution* of the process.

In the finite-dimensional case the resolution (6.26) takes the form

$$z(t) = \sum_{k=1}^{n} c_k e^{i\lambda_k t},$$

where the c_k are independent random variables: $(c_k, c_j) = 0$ for $k \neq j$. This is a representation, of which (6.26) is the continuous analog, of a stationary process in the form of a superposition of uncorrelated stochastic harmonic oscillations.

From the spectral resolution (6.26) one easily obtains the following representation for the correlation function of a stationary process:

$$v(t-s) = \int_{-\infty}^{\infty} e^{i\lambda(t-s)}\, dF(\lambda). \tag{6.27}$$

Here $F(\lambda)$ is a nondecreasing function such that

$$F(\lambda + \Delta\lambda) - F(\lambda) = M\,|\Phi(\lambda, \lambda + \Delta\lambda)|^2. \tag{6.28}$$

Also

$$M\,|z(t)|^2 = \int_{-\infty}^{\infty} dF(\lambda) < \infty.$$

Definition. Let $z(t)$ $(-\infty < t < \infty)$ be a stochastic process whose correlation function $v(t, s) = Mz(t)\overline{z(s)}$ is differentiable. The function

$$w(t, s) = -\left.\frac{\partial v(t+\tau, s+\tau)}{\partial \tau}\right|_{\tau=0} \tag{6.29}$$

will be called the *infinitesimal correlation function* (i.c.f.) of the process $z(t)$. The greatest rank r $(0 \leqslant r \leqslant \infty)$ of all of the quadratic forms

$$\sum_{\alpha,\beta=1}^{n} w(t_\alpha, t_\beta)\,\xi_\alpha\bar{\xi}_\beta \quad (-\infty < t_1, t_2, \ldots, t_n < \infty;\ n = 1, 2, \ldots) \tag{6.30}$$

will be called the *rank of the process* $z(t)$.

Clearly, the i.c.f. $w(t, s)$ and hence the rank r of a stationary process are equal to zero. These quantities therefore characterize the extent to which a process is not stationary.

There arises the question as to what conditions the function $v(t, s)$ must satisfy in order for it to be the correlation function of a linearly representable process with a given rank r and an operator A belonging to a given class of linear operators acting in a Hilbert space H. The process will be stationary if and only if $v(t, s)$ is a positive semidefinite self-adjoint kernel that depends only on the difference of the arguments. The answer to this question for certain classes of nonstationary processes is given in the following chapters.

§3. GAUSSIAN STOCHASTIC PROCESSES

A real process $z(t)$ will be said to be *Gaussian* if the vector $(z(t_1), \dots, z(t_n))$ has a normal distribution for any $t_1, t_2, \dots, t_n \in R_1$ or, equivalently, if

$$Me^{i\sum_k z(t_k)\xi_k} \equiv e^{i\sum_k Mz(t_k) - \frac{1}{2}\sum_{k,l} v(t_k, t_l)\xi_k\xi_l}, \tag{6.31}$$

where the ξ_k $(k = 1, 2, \dots, n)$ are real numbers.

In the particular case when $Mz(t) = 0$ relation (6.31) can be written as follows:

$$Me^{i\sum_k z(t_k)\xi_k} = e^{-\frac{1}{2}\left\|\sum_k \xi_k z(t_k)\right\|^2}. \tag{6.32}$$

where $\|z\| = \sqrt{(z, z)}$, the scalar product being considered in the Hilbert space H_z spanned by the values of $z(t)$.

With the use of relation (6.32) we can introduce the notion of a Gaussian complex process.

A complex process $z(t)$ $(Mz(t) = 0)$ will be called a *Gaussian complex process* if for any $t_1, t_2, \dots, t_n \in R_1$ the stochastic vector $(z(t_1), z(t_2), \dots, z(t_n))$ satisfies the relation

$$Me^{i\,\mathrm{Re}\sum_k z(t_k)\bar{\xi}_k} = e^{-\frac{1}{4}\left\|\sum_k \xi_k z(t_k)\right\|^2} = e^{-\frac{1}{4}\sum_{kl} v(t_k, t_l)\xi_l\bar{\xi}_k}, \tag{6.33}$$

where $\xi_1, \xi_2, \dots, \xi_n$ are complex numbers.

Definition (6.33) implies that in the case of a Gaussian complex process $\mathrm{Re}\, z(t)$ and $\mathrm{Im}\, z(t)$ are independent and equally distributed and $M\,\mathrm{Re}\, z(t)\,\mathrm{Im}\, z(t) = 0$.

We cite without proof the following theorem [18].

Theorem 6.1. *Suppose* $z(t)$ *is a Gaussian complex process and* $Mz(t) = 0$. *Then the correlation function* $v(t, s) = Mz(t)\overline{z(s)} = (z(t), z(s))$ *is a positive semidefinite self-adjoint kernel. Conversely, a given positive semidefinite selfadjoint kernel* $v(t, s)$ *is the correlation function of some Gaussian complex process.*

§4. STOCHASTIC OPEN SYSTEMS

Let F_γ be an open system on an arc γ ($\gamma \in R_n$) in the sense of Chapter III, §1, and let

$$H \dotplus L_2(\gamma, E) \xrightarrow{R} L_2(\gamma, H),$$
$$H \dotplus L_2(\gamma, E) \xrightarrow{S} H \dotplus L_2(\gamma, E) \tag{6.34}$$

be the corresponding input-to-internal-state and input-to-output mappings.

Definition. An open system F_γ is said to be *stochastic* if the spaces H and E are subspaces of a space $L_2(\Omega)$, where Ω is a set of elementary events.

The definition given in Chapter III of a coupling of open systems carries over without change to stochastic open systems. It should be noted in this connection that under a coupling the internal states turn out to be uncorrelated, since they lie in orthogonal subspaces; but these states are at the same time functionally connected through the external space E and the dimension of E can serve as a measure of this connection.

Let us consider a linearly representable field

$$z(x) = e^{i\sum_k A_k x^k} z_0 \quad (z(x) \in H_z \subset L_2(\Omega)) \tag{6.35}$$

and show that it can be included in a stochastic open system associated with some regular family of metric colligations. For this purpose it suffices to consider the system of metrics $2\,\mathrm{Im}\,A_1, 2\,\mathrm{Im}\,A_2, \ldots, 2\,\mathrm{Im}\,A_n$ in H_z and, in accordance with Theorem 1.1, to include this system in the nondegenerate inductor

$$(H_z,\ \varphi,\ E,\ \mu_1,\ \ldots,\ \mu_n), \tag{6.36}$$

where φ is the orthogonal projection onto the subspace

$$E = H_z \ominus \bigcap \operatorname{Rad} \mu_k$$

of H_z (hence $E \subset H_z \subset L_2(\Omega)$) and μ_k $(k = 1, 2, \ldots, n)$ is the restriction of the metric $2 \operatorname{Im} A_k$ to E. But then the aggregate

$$(A_1,\ A_2,\ \ldots,\ A_n,\ H_z,\ \varphi,\ E,\ \mu_1,\ \ldots,\ \mu_n) \tag{6.37}$$

is a vector local colligation.

Now to the colligation (6.37) we can put in correspondence the associated open system whose equations have the form (see §1 of Chapter III)

$$\begin{aligned} & i\frac{dz}{dt} + \left(\sum_{k=1}^{n} A_k \dot{x}^k\right) z = \sum_{k=1}^{n} \varphi_k^{+}[u]\, \dot{x}^k, \\ & z(0) = z_0, \qquad v = u - i\varphi z. \end{aligned} \tag{6.38}$$

If we put $u \equiv 0$, the corresponding internal state will coincide with the given linearly representable stochastic field (6.35).

The inclusion of a linearly representable stochastic field in a stochastic open system raises the possibility of "splitting" a stochastic field into "subfields" if the operators A_k $(k = 1, 2, \ldots, n)$ have a common invariant subspace. To this end we use Theorem 3.4 of Chapter III and its Corollary.

Let $H = H_1 \oplus H_2$, where H_2 is an invariant subspace of the operators A_k $(k = 1, 2, \ldots, n)$. Then (3.35) implies

$$R_\gamma(z_0,\ 0) = R_\gamma^{(1)}(z_0^{(1)},\ 0) + R_\gamma^{(2)}\{z_0^{(2)},\ S^{(1)}(z_0^{(1)},\ 0)\}. \tag{6.39}$$

As follows from (6.38), the expression for $R_\gamma(z_0,\ u)$ has the form

$$z(x^1, \ldots, x^n) = R_\Upsilon(z_0, u) = e^{i\sum_{k=1}^{n} A_k x^k} z(0, 0, \ldots, 0) - i\int_0^Q e^{i\sum_{k=1}^{n} A_k(x^k - y^k)} \sum_{m=1}^{n} \varphi_m^+ [u(Q_y)]\, dy^m, \tag{6.40}$$

where $0 = (0, 0, \ldots, 0)$ and $Q = (x^1, x^2, \ldots, x^n)$.

Since

$$R_\Upsilon^{(1)}(z_0^{(1)}, 0) = e^{i\sum_{k=1}^{n} A_k^{(1)} x^k} z_0^{(1)} \text{ and } S^{(1)}(z_0^{(1)}, 0) = -i\varphi^{(1)} e^{i\sum_{k=1}^{n} A_k^{(1)} x^k} z_0^{(1)},$$

we finally obtain the following formula for splitting the field $e^{i\sum_{k=1}^{n} A_k x^k} z_0$:

$$e^{i\sum_{k=1}^{n} A_k x^k} z_0 = e^{i\sum_{k=1}^{n} A_k^{(1)} x^k} z_0^{(1)} + e^{i\sum_{k=1}^{n} A_k^{(2)} x^k} z_0^{(2)} - \\ - \int_0^Q e^{i\sum_{m=1}^{n} A_m^{(2)}(x^m - y^m)} \sum_{k=1}^{n} \varphi_k^{(2)+} \varphi^{(1)} e^{i\sum_{l=1}^{n} A_l^{(1)} y^l} z_0^{(1)}\, dy^k \tag{6.41}$$

§ 5. QUASIHOMOGENEOUS STOCHASTIC FIELDS

Let $z(x)$ be a linearly representable stochastic field:

$$z(x) = e^{i\sum_{k=1}^{n} A_k x^k} z_0, \tag{6.42}$$

where the A_k are commuting bounded operators in H_z and z_0 is a fixed element of H_z.

The vector i.c.f. of this field is

$$\vec{W}(x, y) = -\nabla_u v(x+u, y+u)/_{u=0}, \tag{6.43}$$

where $v(x, y)$ is the correlation function of $z(x)$.

By virtue of (4.38) the function $\vec{W}(x, y)$ can be represented in the form

$$\vec{W}_k(x, y) = (2\,\mathrm{Im}\, A_k z(x), z(y)). \tag{6.44}$$

Consider the quadratic forms

$$\sum_{\alpha,\beta=1}^{m_k} W_k(x_\alpha, x_\beta)\,\xi_\alpha\bar{\xi}_\beta \quad (x_\alpha, x_\beta \in R_n;\ m_k = 1, 2, \ldots). \tag{6.45}$$

A stochastic field $z(x)$ is said to be *quasihomogeneous* if the rank of the quadratic forms (6.45) is bounded. In the case $n = 1$ the corresponding process will be said to be *quasistationary*.

It follows from (6.44) that $z(x)$ is quasihomogeneous if the ranks of the operators $2\operatorname{Im} A_k$ are finite. In fact, from (4.62) we see that

$$2\operatorname{Im} A_k = \sum_{\gamma=1}^{r_k} \pm(\cdot, g_{k,\gamma})\, g_{k,\gamma} \tag{6.46}$$

and

$$\sum_{\alpha,\beta=1}^{N} \sum_{\gamma=1}^{r_k} \pm (z(x_\alpha), g_{k,\gamma})(g_{k,\gamma}, z(x_\beta))\,\xi_\alpha\bar{\xi}_\beta = $$
$$= \sum_{\gamma=1}^{r_k} \pm \Big|\Big(\sum_{\alpha=1}^{N} z(x_\alpha)\,\xi_\alpha,\ g_{k,\gamma}\Big)\Big|^2. \tag{6.47}$$

Hence the ranks of the quadratic forms do not exceed $\max(r_1, \ldots, r_n)$.

In the case $n = 1$ we have

Theorem 6.2. *In order for a linearly representable process $z(t) = e^{itA} z_0$ to be quasistationary it is necessary and sufficient that the operator A have a finite-dimensional non-Hermitian subspace G_A. Here $r_A = \dim G_A$ coincides with the maximal rank ρ of the quadratic forms*

$$\sum_{\alpha,\beta=1}^{m} w(t_\alpha, t_\beta)\,\xi_\alpha\bar{\xi}_\beta \quad (-\infty < t_1, \ldots, t_m < \infty;\ m = 1, 2, \ldots). \tag{6.48}$$

The sufficiency follows from formulas (6.47), since in this case $\rho \leqslant r_A$. To prove the necessity we note that (6.44) implies

$$(2\operatorname{Im} Az, z) = \sum_{\alpha,\beta=1}^{m} w(t_\alpha, t_\beta)\,\xi_\alpha\bar{\xi}_\beta \quad (z \in H(t_1, \ldots, t_m)), \tag{6.49}$$

where $H(t_1, \ldots, t_m)$ is the subspace consisting of all elements of the form $\sum_{\alpha=1}^{m} \xi_\alpha z(t_\alpha)$ for fixed values $t_1, t_2, \ldots, t_m$.

We introduce in addition the subspace $G(t_1, \ldots, t_m) = P_m (2\,\mathrm{Im}\,A) P_m H$, where P_m is the orthogonal projection onto $H(t_1, \ldots, t_m)$. Clearly, $G(t_1, \ldots, t_m) \subset G_A$. From (6.49) it follows that the rank of

$$\sum_{\alpha, \beta=1}^{m} w(t_\alpha, t_\beta) \xi_\alpha \bar{\xi}_\beta = \dim G(t_1, \ldots, t_m). \tag{6.50}$$

We now choose a countable dense set of points $t_1, t_2, \ldots, t_n, \ldots$ on the real line and consider the sequence of subspaces $H(t_1) \subset H(t_1, t_2) \subset \ldots \subset H(t_1, \ldots, t_m) \subset \ldots$ Since the values of $z(t)\,(-\infty < t < \infty)$ span the whole space H_z, we have $\lim_{m \to \infty} P_m = I$ and hence $\lim_{m \to \infty} G(t_1, \ldots, t_m) = G_A$. On the other hand, (6.50) implies $\dim G(t_1, \ldots, t_m) < \rho$. Therefore $r_A = \dim G_A = \lim_{m \to \infty} \dim G(t_1, \ldots, t_m) \leqslant \rho$. Since $\rho \leqslant r_A$, we conclude that $\rho = r_A$.

CHAPTER VII

DISSIPATIVE PROCESSES OF FINITE RANK

§1. DISSIPATIVE STOCHASTIC PROCESSES

A stochastic process $z(t)$ $(-\infty < t \infty)$ is said to be *dissipative* if all of the quadratic forms (6.30) are positive semidefinite:

$$\sum_{\alpha,\beta=1}^{n} w(t_\alpha, t_\beta)\,\xi_\alpha\bar{\xi}_\beta \geqslant 0. \tag{7.1}$$

If a process $z(t)$ is dissipative, the quadratic forms $\sum_{\alpha,\beta=1}^{n} v(t_\alpha+\tau, t_\beta+\tau)\,\xi_\alpha\bar{\xi}_\beta$ are nonincreasing functions of τ $(-\infty < \tau < \infty)$.

In fact,

$$-\frac{\partial}{\partial\tau}\sum_{\alpha,\beta=1}^{n} v(t_\alpha+\tau, t_\beta+\tau)\,\xi_\alpha\bar{\xi}_\beta = \sum_{\alpha,\beta=1}^{n} w(t_\alpha+\tau, t_\beta+\tau)\,\xi_\alpha\bar{\xi}_\beta \geqslant 0. \tag{7.2}$$

In particular, $v(t, t)$ is a nonincreasing function of t. Inasmuch as in many problems the expression $v(t, t)$ has the physical meaning of mean energy of the process at a given moment of time, condition (7.1) means that energy is dispersed with the passage of time.

It follows from (7.2) that the limit $\lim_{t\to\infty} v(t, t) = \sigma^2_\infty$ exists.

Two cases are possible-

1) $\sigma_\infty = 0$,

2) $\sigma_\infty > 0$.

In the first (second) case the dissipative process $z(t)$ will be said to be *asymptotically damped* (*asymptotically undamped*).

Lemma 7.1. For *a dissipative stochastic process the limit* $\lim_{\tau\to\infty} v(t+\tau, s+\tau)$ *always exists.*

Proof. Consider the relation

$$v(t, s) = (z(t), z(s)) = \frac{1}{4}\{\|z(t)+z(s)\|^2 + \|z(t)-z(s)\|^2 + \\ + i\|z(t)+z(s)\|^2 - i\|z(t)-z(s)\|^2\}. \tag{7.3}$$

Since

$$\left\|\sum_{\alpha=1}^{n} z(t_\alpha+\tau)\xi_\alpha\right\|^2 = \sum_{\alpha,\beta=1}^{n} v(t_\alpha+\tau, t_\beta+\tau)\xi_\alpha\bar{\xi}_\beta,$$

for a dissipative stochastic process the expression

$$\left\|\sum_{\alpha=1}^{n} z(t_\alpha+\tau)\xi_\alpha\right\|^2 \tag{7.4}$$

is a nonincreasing function of τ.

We replace t by $t+\tau$ and s by $s+\tau$ in (7.3). Then each summand in (7.3) has a limit as $\tau\to\infty$.

We note that $\lim_{\tau\to\infty} v(t+\tau, s+\tau)$ is a function depending only on the difference $t-s$:

$$\lim_{\tau\to\infty} v(t+\tau, s+\tau) = v_\infty(t-s) \tag{7.5}$$

Clearly, $v_\infty(t-s)$ is a positive semidefinite self-adjoint kernal and under the assumption of continuity can be regarded as the correlation function of some stationary process.

From the definition of $w(t, s)$ we obtain the following formula for dissipative stochastic processes:

$$v(t, s) = v_\infty(t-s) + \int_0^\infty w(t+\tau, s+\tau)\, d\tau. \tag{7.6}$$

If a process is asymptotically damped, the correlation function can be expressed in terms of the i.c.f. by means of the equality

$$v(t, s) = \int_0^\infty w(t+\tau, s+\tau)\, d\tau. \tag{7.7}$$

In the case of a linearly representable dissipative process

$$z(t) = e^{itA} z_0, \tag{7.8}$$

relation (6.44) implies

$$(2 \operatorname{Im} Az, z) \geqslant 0 \tag{7.9}$$

i.e. the operator A is dissipative.

Definition. A linearly representable dissipative stochastic process $z(t)$ is said to be *complete* if the corresponding operator A is complete.

From the above remarks it follows that a complete dissipative stochastic process is asymptotically damped.

Asymptotically undamped dissipative processes can appear only in connection with the incompleteness of the corresponding operator, i.e. when the real spectrum of A begins to affect substantially the properties of the process.

Remark. The notions of correlation function, i.c.f., linear representability, dissipativeness, etc., introduced above, are meaningful; and the assertions obtained above are clearly valid, not only for stochastic processes but also for curves in an arbitrary Hilbert space. A process $z(t)$, as was shown earlier, can be regarded as a curve in the special Hilbert space $H_z \subset L_2(\Omega)$.

From (6.44) and (6.46) it follows that the i.c.f. of a linearly representable dissipative process of finite rank has the form

$$w(t, s) = \sum_{\alpha=1}^{r} \Phi_\alpha(t)\overline{\Phi_\alpha(s)}, \tag{7.10}$$

where the functions $\Phi_\alpha(t)$ are defined by the relation

$$\Phi_\alpha(t) = (e^{itA}z_0, g_a), \tag{7.11}$$

in which the g_α are channel vectors of A.

There arises the question as to what the functions $\Phi_\alpha(t)$ $(\alpha = 1, 2, \ldots, r)$ must be in order for $w(t, s) = \sum_{\alpha=1}^{r} \Phi_\alpha(t)\overline{\Phi_\alpha(s)}$ to be the i.c.f. of some dissipative process. In the sequel we will elucidate the form and give a construction of the functions $\Phi_\alpha(t)$.

§2. COMPLETE DISSIPATIVE PROCESSES OF RANK ONE

Let

$$z(t) = e^{itA}z_0 \tag{7.12}$$

be a linearly representable process of rank one in which A is a complete dissipative bounded operator with a one-dimensional non-Hermitian subspace.

Then by virtue of (7.10) and (4.34) the i.c.f. of $z(t)$ has the form

$$w(t, s) = \Phi(t)\overline{\Phi(s)}, \tag{7.13}$$

where

$$\Phi(t) = (e^{itA}z_0, g) = -\frac{1}{2\pi i}\oint_\Gamma e^{i\lambda t}((A - \lambda I)^{-1}z_0, g)\,d\lambda =$$
$$-\frac{1}{2\pi i}\oint_\Gamma e^{i\lambda t}(z_0, (A^* - \bar{\lambda} I)^{-1}g)\,d\lambda, \tag{7.14}$$

in which the contour of integration encloses the λ-set $\Lambda = (\lambda_1, \lambda_2, \ldots)$ consisting of the eigenvalues of A.

We can construct a triangular model $\dot{A}$ of A by means of formulas (5.13) and use it to calculate the integral (7.14). Since A and $\dot{A}$ are unitarily equivalent, the scalar product in the integrand of (7.14) does not change in going over from A to $\dot{A}$. Since in this connection z_0 goes over into $Uz_0 = f_0$, where U is a unitary operator, we get

$$\Phi(t) = -\frac{1}{2\pi i}\oint_{\gamma} e^{i\lambda t}(f_0, (\dot{A}^* - \bar{\lambda}I)^{-1}\dot{g})\, d\lambda. \tag{7.15}$$

If we introduce in the space $\dot{H} = l_2$ the basis

$$h_1 (1, 0, 0, \ldots), \ \ldots, \ h_k = (0, 0 \ \ldots, 0, 1, 0, \ldots), \ \ldots,$$

then $\Phi(t)$ can be represented in the form

$$\Phi(t) = -\frac{1}{2\pi i}\oint_{\gamma} e^{i\lambda t}\sum_{k=1}^{\infty} f_{0_k}\overline{\dot{g}_k(\lambda)}\, d\lambda, \tag{7.16}$$

where

$$f_{0_k} = (f_0, h_k), \tag{7.17}$$

$$\dot{g}_k(\lambda) = ((\dot{A}^* - \bar{\lambda}I)^{-1}\dot{g}, h_k). \tag{7.18}$$

The series $\sum_k f_{0_k}\overline{\dot{g}_k(\lambda)}$ converges uniformly on the contour γ, since its remainders satisfy the estimate

$$\sum_N^{\infty} | f_{0_k}\overline{\dot{g}_k(\lambda)} |^2 < \sum_N^{\infty} | f_{0_k} |^2 \cdot \sum_N^{\infty} | \dot{g}_k(\lambda) |^2 < \| \dot{g}(\lambda) \|^2 \sum_N^{\infty} | f_{0_k} |^2. \tag{7.19}$$

But $\| \dot{g}(\lambda) \|$ is uniformly bounded on the contour γ, while $\sum_N^{\infty} | f_{0_k} |^2$ can be made arbitrarily small by taking N sufficiently large.

Changing the order of summation and integration in (7.16), we get

$$\Phi(t)=\sum_{k=1}^{\infty} f_{0k}\left\{-\frac{1}{2\pi i}\oint_{\gamma} e^{i\lambda t}\overline{g_k(\lambda)}\,d\lambda\right\} \tag{7.20}$$

while formula (5.34) implies

$$\overline{g_k(\lambda)}=\sqrt{2\operatorname{Im}\lambda_k}\,\frac{1}{\lambda_k-\lambda}\prod_{j=1}^{k-1}\frac{\bar{\lambda}_j-\lambda}{\lambda_j-\lambda}. \tag{7.21}$$

Consequently,

$$\Phi(t)=\sum_{k=1}^{\infty} f_{0k}\Lambda_k(t), \tag{7.22}$$

where

$$\Lambda_k(t)=-\frac{1}{2\pi i}\sqrt{2\operatorname{Im}\lambda_k}\oint_{\gamma} e^{i\lambda t}\frac{1}{\lambda_k-\lambda}\prod_{j=1}^{k-1}\frac{\bar{\lambda}_j-\lambda}{\lambda_j-\lambda}\,d\lambda. \tag{7.23}$$

From (7.23) we see that the functions $\Lambda_k(t)$ determining the structure of $\Phi(t)$ are uniquely determined by the λ-set. In the sequel the functions defined by relation (7.23) will be called the *special λ-functions.*

Applying the theory of residues to (7.23), we see that the special λ-functions can be represented in the form

$$\Lambda_k(t)=\sum_{j=1}^{k} P_{k,j}(t)\,e^{i\lambda_j t}, \tag{7.24}$$

where $P_{k,j}(t)$ is a polynomial of degree less than the multiplicity of the eigenvalue λ_j.

Thus we have proved

Theorem 7.1. *If $z(t)$ is a complete dissipative process of rank one, its i.c.f. has the form*

$$w(t,s)=\Phi(t)\overline{\Phi(s)}, \tag{7.25}$$

where

$$\Phi(t) = \sum_{k=1}^{\infty} c_k \Lambda_k(t), \tag{7.26}$$

$$\sum_{k=1}^{\infty} |c_k|^2 < \infty, \tag{7.27}$$

and the $\Lambda_k(t)$ *are the special* λ*-functions* (7.23).

The correlation function of $z(t)$ is determined (by virtue of (7.7) and (7.25)) from the relation

$$v(t, s) = \int_0^{\infty} \Phi(t+\tau)\overline{\Phi(s+\tau)}\, d\tau. \tag{7.28}$$

Theorem 7.2. *Let* $\{\lambda_1, \lambda_2, \ldots\}$ *be an arbitrary* λ*-set and let* $c_1, c_2, \ldots$ *be an arbitrary sequence of complex numbers satisfying the condition* $\sum_{k=1}^{\infty} |c_k|^2 < \infty$. *Then there exists a complete linearly representable dissipative Gaussian process* $\mathbf{z}(t)$ *of rank one whose i.c.f. has the form*

$$w(t, s) = \Phi(t)\overline{\Phi(s)},$$

where

$$\Phi(t) = \sum_{k=1}^{\infty} c_k \Lambda_k(t),$$

and whose correlation function is given by expression (7.28).

Proof. We construct with respect to the given λ-set a triangular model of form (5.13) and, letting $f_0 = (c_1, c_2, \ldots)$, consider the curve

$$f(t) = e^{it\dot{A}} f_0 \tag{7.29}$$

in l_2.

This curve is by construction dissipative and asymptotically damped. The operator $\dot{A}$ is complete while the l.c.f. of $f(t)$, as follows from formulas (6.44), has the form (7.25) with $\Phi(t)$ being of form (7.26). The correlation function of $f(t)$ is given by relation (7.28).

The function $v(t, s) = (f(t), f(s))$ is a continuous positive semidefinite self-adjoint kernel. By virtue of Theorem 6.1 there exists a Gaussian complex process $z(t)$ whose correlation function coincides with $v(t, s) = (f(t), f(s))$. Since the correlation functions of $f(t)$ and $z(t)$ coincide, it follows by virtue of Lemma 6.2 that the process $z(t)$ admits the representation $z(t) = e^{itA}z_0$, where A is unitarily equivalent to the operator $\dot{A}$ of form (5.13).

§3. FINITE-DIMENSIONAL DISSIPATIVE PROCESSES OF RANK ONE

An especially simple form is achieved by the correlation function of a linearly representable dissipative process $z(t) = e^{itA}z_0$ of rank one under the assumption that the space H_z spanned by the values of $z(t)$ $(-\infty < t < \infty)$ is finite-dimensional and all of the nonreal eigenvalues $\lambda_1, \lambda_2, \ldots, \lambda_n$ of A are distinct. The space H_z can be represented in the form of a sum $H_z^0 \oplus H_z'$ of invariant subspaces of A such that A is self-adjoint on H_z^0 but completely non-self-adjoint on H_z'. Then the correlation function $v(t, s)$ will have the form $v(t, s) = v^0(t, s) + v'(t, s)$, where $v^0(t, s) = \sum_{k=1}^{m} \rho_k e^{i\mu_k(t-s)}$ $(\rho_k > 0, -\infty < \mu_k < \infty,\ m \leqslant n)$ is the correlation function of a stationary process $e^{itA}z^0$ $(z^0 \in H_z^0)$.

It but remains to find the form of the function $v'(t, s)$. We can therefore assume without loss of generality that $H_z' = H_z$.

Theorem 7.3. *In order for a given function $v(t, s)$ $(-\infty < t, s < \infty)$ to be the correlation function of some finite-dimensional linearly representable dissipative process of rank one with a simple nonreal spectrum it is necessary and sufficient that it admit a representation of the form*

$$v(t, s) = \sum_{k=1}^{m} \rho_k e^{i\mu_k(t-s)} + i \sum_{k, j=1}^{p} \frac{c_k \bar{c}_j}{\lambda_k - \bar{\lambda}_j} e^{i(\lambda_k t - \bar{\lambda}_j s)}, \tag{7.30}$$

where the c_k $(k = 1, 2, \ldots, n)$ are arbitrary numbers, $\operatorname{Im} \lambda_k > 0$ *and $\lambda_k \neq \lambda_j$ for $k \neq j$.*

The i.c.f. in this case has the form

$$w(t, s) = \Phi(t)\overline{\Phi(s)}, \tag{7.31}$$

where

$$\Phi(t) = \sum_{k=1}^{k} c_k e^{i\lambda_k t}. \tag{7.32}$$

Proof. Suppose $H_z' = H_z$. Let $z(t) = e^{itA} z_0$ be a process satisfying the conditions of Theorem 7.1 and let $z_1, z_2, \ldots, z_n$ be eigenvectors of A corresponding to the eigenvalues $\lambda_1, \ldots, \lambda_n$. Representing z_0 in the form

$$z_0 = \sum_{k=1}^{n} a_k z_k, \tag{7.33}$$

we get

$$e^{itA} z_0 = \sum_{k=1}^{n} a_k e^{i\lambda_k t} z_k. \tag{7.34}$$

Since $r = 1$, it follows by virtue of (7.10) that

$$w(t, s) = \Phi(t)\overline{\Phi(s)}, \tag{7.35}$$

where

$$\Phi(t) = (e^{itA} z_0, g) = \sum_{k=1}^{n} a_k (z_k, g) e^{i\lambda_k t}. \tag{7.36}$$

Setting $c_k = a_k (z_k, g)$, we obtain (7.32).

Further, since

$$v(t, s) = \int_0^\infty w(t + \tau, s + \tau)\, d\tau,$$

we obtain (7.30) by integrating with respect to τ.

Now suppose given a function $v(t, s)$ of the form

$$v(t, s) = i \sum_{k, j=1}^{n} \frac{c_k \bar{c}_j}{\lambda_k - \bar{\lambda}_j} e^{i(\lambda_k t - \bar{\lambda}_j s)}, \tag{7.37}$$

where the c_k $(k = 1, 2, \ldots, n)$ are arbitrary numbers and λ_k are numbers satisfying the conditions $\operatorname{Im} \lambda_k > 0$ and $\lambda_k \neq \lambda_j$ for $k \neq j$.

Clearly,

$$\dot{v}(t, s) = \int_0^\infty w(t + \tau, s + \tau)\, d\tau = \int_0^\infty \Phi(t + \tau)\overline{\Phi(s + \tau)}\, d\tau\,, \qquad (7.38)$$

where

$$\Phi(t) = \sum_{k=1}^{n} c_k e^{i\lambda_k t}. \qquad (7.39)$$

Since

$$\sum_{j,\,k=1}^{n} v(t_j, t_k)\,\xi_j \bar{\xi}_k = \int_0^\infty \Big|\sum_{j=1}^{n} \Phi(t_j + \tau)\,\xi_j\Big|^2 d\tau \geqslant 0, \qquad (7.40)$$

$v(t, s)$ is the correlation function of some Gaussian process $z(t)$. Let us prove that $z(t)$ is a linearly representable dissipative process in a finite-dimensional space. To this end we construct with respect to the numbers $\lambda_1, \ldots, \lambda_n$ a triangular model $\dot{X} = (\dot{A}, \dot{H}, \dot{g}, 1)$ of form (5.13) in which $\dim \dot{H} = n < \infty$. Clearly, the correlation function $\dot{v}(t, s)$ of $e^{it\dot{A}} f_0$, $(f_0 \in \dot{H})$ has the form (7.38), in which the arbitrary numbers c_k $(k = 1, 2, \ldots, n)$ can be obtained by properly choosing f_0. Since $\dot{v}(t, s) = v(t, s)$, we conclude on the basis of Lemma 6.2 that the process $z(t)$ is linearly representable and satisfies all of the requirements of the theorem.

§4. COMPLETE DISSIPATIVE PROCESSES OF FINITE RANK

Theorem 7.4. *If* $z(t)$ *is a complete linearly representable dissipative process of rank* r $(0 < r < \infty)$, *its i.c.f. has the form*

$$w(t, s) = \sum_{\alpha=1}^{r} \Phi_\alpha(t)\overline{\Phi_\alpha(s)}, \qquad (7.41)$$

where

$$\Phi_\alpha(t) = \sum_{k=1}^{\infty} c_{k,\alpha} \Lambda_k(t), \tag{7.42}$$

in which

$$\sum_{k=1}^{\infty} |c_{k,\alpha}|^2 < \infty \tag{7.43}$$

and the $\Lambda_k(t)$ are special λ-functions.

The correlation function of $z(t)$ has the form

$$v(t, s) = \sum_{\alpha=1}^{r} \int_0^\infty \Phi_\alpha(t+\tau) \overline{\Phi_\alpha(s+\tau)}\, d\tau. \tag{7.44}$$

Proof. Since $z(t)$ is a linearly representable process,

$$z(t) = e^{itA} z_0, \tag{7.45}$$

where A is a complete dissipative operator acting in the space H_z. We include A in a complex $(A, H, g_1, \ldots, g_r, I)$ belonging to the class $D_r(\Lambda(A))$.

We construct with respect to the eigenvalues of A, taking into account their multiplicities, a λ-set $\{\lambda_1, \lambda_2, \ldots\}$ and a universal complex $\Theta_r(\Lambda)$ of class D_r.

Consider a subspace $\tilde{H}$ of $\dot{H}$ for which, according to Theorem 5.11 on a universal complex, the projection

$$P_{\tilde{H}}(\dot{A}, \dot{H}, \dot{g}_1, \ldots, \dot{g}_r\, I) = (\tilde{A}, \tilde{H}, \tilde{g}_1, \ldots, \tilde{g}_r, I) \tag{7.46}$$

is a complex that is unitarily equivalent to the complex $(A, H_r, g_1, \ldots, g_r, I)$. By virtue of the unitary equivalence and a formula for the i.c.f. we have

$$\begin{aligned} w(t, s) &= \sum_{\alpha=1}^{r} (e^{itA} z_0, g_\alpha) \overline{(e^{isA} z_0, g_\alpha)} = \\ &\sum_{\alpha=1}^{r} (e^{it\tilde{A}} \tilde{z}_0, \tilde{g}_\alpha) \overline{(e^{is\tilde{A}} \tilde{z}_0, \tilde{g}_\alpha)}. \end{aligned} \tag{7.47}$$

Since $e^{it\tilde{A}}\tilde{z}_0 = e^{it\dot{A}}\tilde{z}_0$ and $e^{it\tilde{A}}\tilde{z}_0 \in \tilde{H}$, while $\tilde{g}_\alpha = P_{\tilde{H}}\dot{g}_\alpha$,

$$w(t, s) = \sum_{\alpha=1}^{r} (e^{it\dot{A}}\tilde{z}_0, \dot{g}_\alpha)\overline{(e^{is\dot{A}}\tilde{z}_0, \dot{g}_\alpha)}. \tag{7.48}$$

From the condition $\dot{g}_\alpha \in l_2^{(\alpha)}$ and the equality

$$e^{it\dot{A}}\tilde{z}_0 = e^{it\dot{A}_1}\tilde{z}_{0_1} + \cdots + e^{it\dot{A}_r}\tilde{z}_{0_r},$$

where $\tilde{z}_{0_\alpha} \in l_2^{(\alpha)}$, it follows that

$$w(t, s) = \sum_{\alpha=1}^{r} (e^{it\dot{A}_\alpha}\tilde{z}_{0_\alpha}, \dot{g}_\alpha)\overline{(e^{is\dot{A}_\alpha}\tilde{z}_{0_\alpha}, \dot{g}_\alpha)}. \tag{7.49}$$

Inasmuch as the $\dot{A}_\alpha$ are operators with a one-dimensional non-Hermitian subspace, (7.10) implies the required representation (7.41) for $w(t, s)$. Equation (7.44) follows from formula (7.7).

Theorem 7.5. *If an arbitrary* λ-set $\{\lambda_1, \lambda_2, \ldots\}$ *and* r *arbitrary sequences* $\{c_{1,\alpha}, c_{2,\alpha}, \ldots, c_{n,\alpha}, \ldots\}$ *of complex numbers satisfying the condition* $\sum_{k=1}^{\infty} |c_{k,\alpha}|^2 < \infty$ $(\alpha = 1, 2, \ldots, r)$ *are given, then there exists a complete linearly representable dissipative Gaussian process* $z(t)$ *of rank not exceeding* r *whose i.c.f. has the form* (7.41), *in which the* $\Phi_\alpha(t)$ *have the form* (7.42), *and whose correlation function* $v(t, s)$ *is given by relation* (7.44).

Proof. With respect to the given λ- set we again construct the universal model complex $\Theta_r(\Lambda)$ defined by equality (5.93).

Consider the process

$$\tilde{z}(t) = e^{itA}\tilde{z}_0, \tag{7.50}$$

where $\tilde{z}_0 = \tilde{z}_{0_1} + \cdots + \tilde{z}_{0_r}$, in which $\tilde{z}_{0_\alpha} = \{c_{k,\alpha}\} \in l_2^{(\alpha)}$, and the corresponding subspace $\tilde{H}_z$ spanned by by the values of $z(t)$.

The i.c.f. of $\tilde{z}(t)$ has the form (see the proof of the preceding theorem)

$$\tilde{w}(t, s) = \sum_{\alpha=1}^{r} (e^{it\dot{A}}\tilde{z}_0, \dot{g}_\alpha)\overline{(e^{is\dot{A}}\tilde{z}_0, \dot{g}_\alpha)} =$$

$$\sum_{\alpha=1}^{r} (e^{it\dot{A}_\alpha}\tilde{z}_{0_\alpha}, \dot{g}_\alpha)\overline{(e^{is\dot{A}_\alpha}\tilde{z}_{0_\alpha}, \dot{g}_\alpha)} = \sum_{\alpha=1}^{r} \Phi_\alpha(t)\overline{\Phi_\alpha(s)}, \tag{7.51}$$

where by virtue of relation (7.11) the $\Phi_\alpha(t)$ are the functions (7.42) given in the conditions of the theorem. Hence $\tilde{w}(t, s) = w(t, s)$ and consequently $(\tilde{z}(t), \tilde{z}(s)) = \tilde{v}(t, s) = v(t, s) = (z(t), z(s))$. But, as in Theorem 7.3, this implies the existence and linear representability of the process $z(t)$.

We note that Theorem 7.3 can also be extended to finite-dimensional processes of any finite rank r $(1 \leqslant r < \infty)$. It is only necessary to replace the numbers c_k (c_k), $\bar{c}_k$ $(k = 1, 2, \ldots, n)$ in formula (7.30) by the row (column) matrices $c_k = \| c_{k_1}, \ldots, c_{k_r} \|$.

$$c_k^* = \left\| \begin{matrix} \bar{c}_{k_1} \\ \vdots \\ \bar{c}_{k_r} \end{matrix} \right\| .$$

§5. DISSIPATIVE PROCESSES OF RANK ONE WITH SPECTRUM AT THE ORIGIN

Theorem 7.6. *Suppose*

$$z(t) = e^{itA} z_0, \tag{7.52}$$

is a linearly representable dissipative process of rank one whose spectrum consists of the single point $\lambda = 0$. *Then the i.c.f. of* $z(t)$ *has the form*

$$w(t, s) = \Phi(t) \overline{\Phi(s)}, \tag{7.53}$$

where

$$\Phi(t) = \int_0^t J_0\left(2\sqrt{t\xi}\right) f_0(\xi)\, d\xi, \tag{7.54}$$

in which $J_0(\cdot)$ *is the Bessel function of the first kind of zero order. In addition, the process* $z(t)$ *is asymptotically damped and its correlation function has the form*

$$v(t, s) = \int_0^\infty \Phi(t+\tau) \overline{\Phi(s+\tau)}\, d\tau. \tag{7.55}$$

Proof. We construct a triangular model of the form $\dot{X} = (\dot{A}, L_2(0, l), \dot{g} = 1, J = 1)$ with the use of (5.36), where in the present case $\alpha(x) \equiv 0$ and $\dot{A}$ is defined by the relation

$$\dot{A}f = i\int_x^l f(\xi)\,d\xi. \tag{7.56}$$

We make use of formula (7.10), which in this case has the form

$$w(t, s) = \Phi(t)\overline{\Phi(s)}, \tag{7.57}$$

where

$$\Phi(t) = -\frac{1}{2\pi i}\oint_\gamma ((\dot{A} - \lambda I)^{-1} f_0, 1)\, e^{it\lambda}\, d\lambda, \tag{7.58}$$

$$f_0 = f_0(\xi) \quad (0 \leqslant \xi \leqslant l). \tag{7.59}$$

From relation (7.56) it follows that $\Phi(t)$ can be represented in the form

$$\Phi(t) = \int_0^l f_0(\xi)\left[\frac{1}{2\pi i}\oint_\gamma \frac{e^{i\left(t\lambda + \frac{\xi}{\lambda}\right)}}{\lambda}\,d\lambda\right] d\xi. \tag{7.60}$$

Calculating this integral by means of the theory of residues, we finally get

$$\Phi(t) = \int_0^l f_0(\xi)\left[\frac{1}{2\pi i}\oint_\gamma \frac{1}{\lambda}\sum_{n=0}^{\infty} \frac{(-1)^n t^n \xi^n}{(n!)^2}\,d\lambda\right] d\xi =$$
$$\int_0^l f_0(\xi)\sum_{n=0}^{\infty}\frac{(-1)^n (t\xi)^n}{(n!)^2}\,d\xi = \int_0^l f_0(\xi)\, J_0\left(2\sqrt{t\xi}\right) d\xi. \tag{7.61}$$

Thus $\Phi(t)$ has the form (7.54). The asymptotic damping follows from Theorem 5.13, and then the correlation function is recovered from the i.c.f. by formula (7.55).

Theorem 7.7. *For any given function* $f_0(x) \in L_2(0, l)$ *there exists an asymptotically damped linearly representable dissipative Gaussian process whose spectrum is concentrated at the origin and whose i.c.f. has the form* (7.53), *in which* $\Phi(t)$ *is the function defined by relation* (7.54).

The proof of this theorem is carried out analogously to the proof of Theorem 7.2, the only difference being that the operator $\dot{A}$ is defined by equality (7.56) while the linearly representable curve $f(x, t)$ has the form

$$f(x, t) = e^{it\dot{A}} f_0(x). \tag{7.62}$$

§ 6. DISSIPATIVE PROCESSES OF FINITE RANK WITH SPECTRUM AT THE ORIGIN

Theorem 7.8. *A linearly representable dissipative process* $z(t)$ *of rank* r *with spectrum at the origin is asymptotically damped and has an i.c.f. of the form*

$$w(t, s) = \sum_{\alpha=1}^{r} \Phi_\alpha(t)\overline{\Phi_\alpha(s)}, \tag{7.63}$$

where

$$\Phi_\alpha(t) = \int_0^l f_{0,\alpha}(\xi) J_0(2\sqrt{t\xi})\, d\xi$$
$$(f_{0,\alpha}(\xi) \in L_2(0, l)). \tag{7.64}$$

The proof of this theorem can be carried out analogously to the proof of Theorem 7.4 if one chooses $\Theta_r(l)$ (Chapter V, § 5) as a universal complex, includes the operator A in a complex $X = (A, H, g_1, \ldots, g_r, l)$ of class $C_r(l)$ and makes use of Theorem 5.12.

Theorem 7.9. *For any given functions* $f_{0,\alpha}(\xi) \in L_2(0, l)$ $(\alpha = 1, \ldots, r)$ *there exists an asymptotically damped linearly representable dissipative process with spectrum at the origin whose i.c.f. has the form* (7.63), *in which the* $\Phi_\alpha(t)$ *are of the form* (7.64).

The proof of this theorem can be obtained by repeating the arguments presented in the proof of Theorem 7.5 if one chooses $\Theta_r(l)$ as a universal complex and puts $\tilde{z}_{0\alpha} = f_{0,\alpha}(\xi)$.

CHAPTER VIII

SPECTRAL RESOLUTIONS OF NONSTATIONARY PROCESSES

§1. COMPLETE PROCESSES OF FINITE RANK

Consider a stochastic open system associated with a local colligation (A, H, φ, E, μ), where μ is a J-metric. The local colligation condition has the form

$$\frac{1}{i}(A - A^*) = \varphi^* J \varphi, \tag{8.1}$$

where E and H are subspaces of a space $L_2(\Omega)$, the subspace E being the range of the operator $(2 \operatorname{Im} A) H$:

$$E = (2 \operatorname{Im} A) H, \quad \dim E < \infty.$$

Since the operator $2 \operatorname{Im} A$ is self-adjoint in E, it can be represented in the form

$$2 \operatorname{Im} A = \sum_{\alpha=1}^{r} \omega_\alpha (\cdot, a_\alpha) a_\alpha, \tag{8.2}$$

where the a_α form an orthonormal system of eigenvectors of $2\,\mathrm{Im}\,A$ and the ω_α are the corresponding eigenvalues.

Suppose $\omega_1, \omega_2, \ldots, \omega_p > 0$, where $\omega_{p+1}, \ldots, \omega_{p+q} < 0$.

We let

$$g_\alpha = \sqrt{|\omega_\alpha|}\,a_\alpha, \tag{8.3}$$

and take as J the matrix

$$J = \begin{Vmatrix} I_p & 0 \\ 0 & -I_q \end{Vmatrix}, \tag{8.4}$$

where I_p and I_q are the unit matrices of orders p and q respectively. Then (8.2) can be represented in the form

$$2\,\mathrm{Im}\,A = \sum_{\alpha=1}^{r} \varepsilon_\alpha (\cdot, g_\alpha)\, g_\alpha, \tag{8.5}$$

where

$$\varepsilon_\alpha = \begin{cases} 1, & \alpha = 1, 2, \ldots, p; \\ -1, & \alpha = p+1, \ldots, p+q. \end{cases}$$

Thus to the operator colligation (A, H, φ, E, μ) written relative to a basis a_α $(\alpha = 1, 2, \ldots, r)$ in E there corresponds the operator complex

$$\left(A, H, g_1, \ldots, g_r, \; J = \begin{Vmatrix} I_p & 0 \\ 0 & -I_q \end{Vmatrix}\right), \tag{8.6}$$

with

$$\mu(u, v) = \sum_{\alpha=1}^{p} u_\alpha \bar{v}_\alpha - \sum_{\alpha=p+1}^{r} u_\alpha \bar{v}_\alpha, \quad \left(u = \sum_{\alpha=1}^{r} u_\alpha a_\alpha, \quad v = \sum_{\alpha=1}^{r} v_\alpha a_\alpha\right).$$

According to (3.45) and (3.46), the equations of the open system associated with the colligation (A, H, φ, E, μ) have the form

$$\begin{aligned} i\frac{dz}{dt} + Az &= \varphi^{+}[u], \\ z|_{t=0} &= z_0, \end{aligned} \tag{8.7}$$

$$v = u - i\varphi z, \quad z(t) \in H. \tag{8.8}$$

We take as φ the mapping

$$\varphi z = \sum_{\alpha=1}^{r} (z, g_\alpha)\, a_\alpha. \tag{8.9}$$

Then

$$\varphi^+ u = \sum_{\alpha=1}^{r} \varepsilon_\alpha (u, a_\alpha)\, g_\alpha. \tag{8.10}$$

In fact,

$$(\varphi^+ u, z) = \sum_{\alpha=1}^{r} \varepsilon_\alpha (u, a_\alpha)(g_\alpha, z),$$

$$\mu(u, \varphi z) = \sum_{\alpha=1}^{r} \varepsilon_\alpha \overline{(z, g_\alpha)}\,(u, a_\alpha) = \sum_{\alpha=1}^{r} \varepsilon_\alpha (u, a_\alpha)(g_\alpha, z) = (\varphi^+ u, z).$$

We verify the local colligation condition $\varphi^+\varphi = 2\,\mathrm{Im}\,A$:

$$\varphi^+\varphi z = \sum_{\alpha=1}^{1} \varepsilon_\alpha (z, g_\alpha)\, g_\alpha = 2\,\mathrm{Im}\,Az.$$

Setting $(u, a_\alpha) = u_\alpha(t)$, we can now write the open system equations (8.7) and (8.8) in the form

$$i\frac{dz}{dt} + Az = \sum_{\alpha=1}^{r} \varepsilon_\alpha u_\alpha(t)\, g_\alpha,$$
$$z|_{t=0} = z_0, \tag{8.11}$$

$$v_\alpha(t) = u_\alpha(t) - i(z, g_\alpha), \tag{8.12}$$

where $v_\alpha(t) = (v, a_\alpha)$.

Any process $u(t)$ whose values belong to E will be called a *channel process.*

We note that the channel vectors g_α $(\alpha = 1, 2, \ldots, r)$ constitute in the present case a system of uncorrelated random variables $(Mg_\alpha \bar{g}_\beta = 0,\ \alpha \neq \beta)$, with

$$M|g_\alpha|^2 = |\omega_\alpha|. \tag{8.13}$$

A channel process of the form $u'(t) = \sum_{\alpha=1}^{p} u_\alpha(t) a_\alpha$ will be called a *direct process,* while a process of the form $u''(t) = \sum_{\alpha=p+1}^{r=p+q} u_\alpha(t) a_\alpha$ will be called an inverse process.

Theorem 8.1. *The i.c.f.* $w(t, s)$ *of an internal process* $z(t)$ *has the form*

$$w(t, s) = -\frac{\partial}{\partial\tau} v(t+\tau, s+\tau)|_{\tau=0} = K_{v'}(t, s) - K_{u'}(t, s) - [K_{v''}(t, s) - K_{u''}(t, s)], \tag{8.14}$$

in which $K_{u'}(t, s)$ *and* $K_{u''}(t, s)$ *are the correlation functions of the direct and inverse input processes* $u'(t)$ *and* $u''(t)$ *respectively, while* $K_{v'}(t, s)$ *and* $K_{v''}(t, s)$ *are the correlation functions of the direct and inverse output processes* $v'(t)$ *and* $v''(t)$ *respectively.*

Proof. From equation (8.7) and condition (8.1) we easily obtain the relation

$$-w(t, s) + (\varphi^+\varphi z(t), z(s)) = \left(\frac{1}{i}\varphi^+ u(t), z(s)\right) + \left(z(t), \frac{1}{i}\varphi^+ u(s)\right).$$

Making use of the expression for $v(t)$, we get

$$-w(t, s) + \mu(v(t) - u(t), v(s) - u(s)) = \frac{1}{i}\mu(u(t), \varphi z(s)) - \frac{1}{i}\mu(\varphi z(t), u(s)),$$

$$w(t, s) + \mu(v(t), v(s)) - \mu(u(t), v(s)) + \mu(u(t), u(s)) - \mu(v(t), u(s)) = -\mu(u(t), v(s)) + \mu(u(t), u(s)) - \mu(v(t), u(s)) + \mu(u(t), u(s)).$$

Consequently,

$$w(t, s) = \mu(v(t), v(s)) - \mu(u(t), u(s)), \tag{8.15}$$

which implies (8.14).

Thus *the i.c.f. of an internal process is equal to the increase in the correlation function of the channel output process over that of the channel input process, the correlation functions of the direct and inverse processes being taken with "plus" and "minus" signs respectively.*

Corollary. *In the case of a dissipative open system the i.c.f. is equal to the increase*

$$w(t, s) = K_v(t, s) - K_u(t, s) \tag{8.16}$$

in the correlation function of the channel output process over that of the channel input process.

In particular, when the input $u(t)$ is equal to zero, we have

$$w(t, s) = K_v(t, s), \tag{8.17}$$

i.e. the i.c.f. of the internal process coincides with the correlation function of the channel output process.

Let us agree to call a stochastic process $z(t)$ *determinate* if it has the form

$$z(t) = \psi(t) z \tag{8.18}$$

where $\psi(t)$ is a complex-valued function and z is a fixed random variable.

Theorem 8.2. *If $z(t) = e^{itA} z_0$ is a complete dissipative process of finite rank, there exist two sequences of "elementary" stochastic processes $z_k(t)$ and $u_k(t)$ $(k = 1, 2, \dots)$ satisfying the following conditions.*

1. *The processes $z_k(t) = \psi_k(t) z_k$ $(z_k \in H_z)$ are determinate and $Mz_k \bar{z}_j = \delta_{kj}$.*

2. *The following expansion holds:*

$$z(t) = \sum_{k=1}^{\infty} \psi_k(t) z_k. \tag{8.19}$$

3. *The processes $u_k(t)$ $(k = 1, 2, \dots)$ are channel processes, i.e. their values belong to a fixed r dimensional space $E = (2 \operatorname{Im} A) H$:*

$$u_k(t) = \sum_{\alpha=1}^{r} u_{k,\alpha}(t) a_\alpha \quad (u_{k,\alpha}(t) = Mu_k(t) \bar{a}_\alpha), \tag{8.20}$$

where $Ma_\alpha \bar{a}_\beta = \delta_{\alpha\beta}$ and $a_\alpha \in E$.

4. *The functions* $\psi_k(t)$ *and* $u_{k,\alpha}(t)$ *can be determined from the system of recursion equations*

$$i\frac{d\psi_k}{dt}+\lambda_k\psi_k=\sum_{\alpha=1}^{r}u_{k,\alpha}(t)\sqrt{\omega_\alpha}M(a_\alpha\bar{z}_\alpha), \tag{8.21}$$

$$\psi_k(t)|_{t=0}=\psi_k(0), \tag{8.22}$$

$$u_{k+1,\alpha}(t)=u_{k,\alpha}(t)-i\sqrt{\omega_\alpha}M(z_\alpha\bar{a})\psi_k(t) \quad (k=1, 2, \dots), \tag{8.23}$$

$$u_{1,\alpha}(t)=0 \quad (\alpha=1, 2, \dots, r), \tag{8.24}$$

where the λ_k $(k=1, 2, \dots)$ *are the eigenvalues of the operator* A *and the* ω_α $(\alpha=1, 2, \dots, r)$ *are the eigenvalues of* $2\operatorname{Im}A$.

Proof. We include A in a complex $X=(A, H, g_1, \dots, g_r, I)$. By virtue of Lemma 5.3 the operator A^* also is complete. For A^* there exists an increasing system of finite-dimensional invariant subspaces $H_0^*=0\subset H_1^*\subset H_2^*\subset \dots$ such that $\dim(H_{k+1}^*\ominus H_k^*)=1$ and $\lim_{k\to\infty}H_k^*=H$. The subspaces $H_k = H\ominus H_k^*$ $(k=1, 2, \dots)$ clearly form a decreasing sequence $H=H_0\supset H_1\supset H_2\supset\dots$ of invariant subspaces of A such that $\dim(H_{k-1}\ominus H_k) = 1$ and $\lim_{k\to\infty}H_k=0$.

According to Theorem 1.6, the complex X can be represented in the form of a product $X=X_1^\perp\vee X_2^\perp\vee\dots$ where the complex $X_k^\perp$ is the projection of the complex X onto the subspace $H_k^\perp=H_{k-1}\ominus H_k$, $(k=1, 2, \dots)$. Since $\dim H_k^\perp=1$, the complex $X_k^\perp$ $(k=1, 2, \dots)$ is elementary (Chapter IV, §5). Let z_k denote the unit vector of $H_k^\perp$. The complex $X_k^\perp$ has the form

$$X_k^\perp=P_k^\perp X=(A_k^\perp, H_k^\perp, g_1^{(k)}, \dots, g_r^{(k)}, I), \tag{8.25}$$

in which

$$A_k^\perp h_k=\lambda_k h_k, \tag{8.26}$$

$$g_\alpha^{(k)} = P_k^\perp g_\alpha = (g_\alpha, z_k)\, z_k \quad (k = 1, 2, \ldots). \tag{8.27}$$

We set $z_k(t) = \psi_k z_k$, where ψ_k is a number.

By virtue of (6.38) the equations of the stochastic open system $F_k^\perp$ associated with the complex $X_k^\perp$ can be written in the form

$$i \frac{d\psi_k}{dt} + \lambda_k \psi_k = \sum_{\alpha=1}^{r} u_{\alpha, k}(t)\, (g_\alpha, z_k); \tag{8.28}$$

$$\psi_k|_{t=0} = \psi_{k,0};$$
$$v_{\alpha, k}(t) = u_{\alpha, k}(t) - i\, \overline{(g_\alpha, z_k)}\, \psi_k(t). \tag{8.29}$$

Since $z(t) = e^{itA} z_0$ is the internal state of an open system F of form (8.28) when $u(t) \equiv 0$, by making use of the Corollary to Theorem 3.4 on the decomposition of the system F associated with X into a coupling $F = F_1^\perp \vee F_2^\perp \vee \vee \cdots$, we can obtain all of the assertions of Theorem 8.2 if we take into account the fact that for stochastic open systems the scalar product (g_α, z_k) is equal to the mathematical expectation $M g_\alpha \bar{z}_k$ and $g_\alpha = \sqrt{\omega_\alpha} a_\alpha$ by virtue of (8.3).

We note that relations (8.28) and (8.29) are recursion relations whose character can be depicted in the form of a diagram (Fig. 12).

From relations (8.28) and (8.29) or from the diagram we see that the pairwise uncorrelated processes $z_k(t)$ are interconnected by means of the channel processes $u_k(t)$ $(k = 1, 2, \ldots)$. In particular, when $u_1(t) \equiv 0$ we obtain a decomposition of a linearly representable process; the elementary links in this decomposition turn out to be interconnected by means of stochastic (channel) processes lying in a fixed space whose dimension coincides with the rank of the process (Fig. 13).

The expansion (8.19) is a generalization of the corresponding spectral resolution of a stationary process, since when $r = 0$ the space E and the elementary processes $z_k(t)$ are not interconnected.

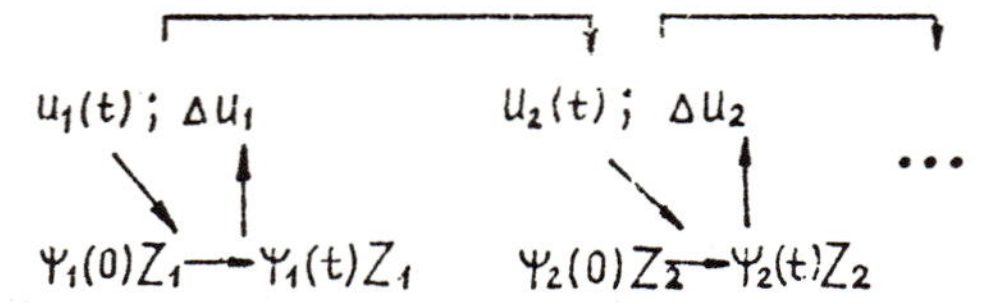

Fig. 12

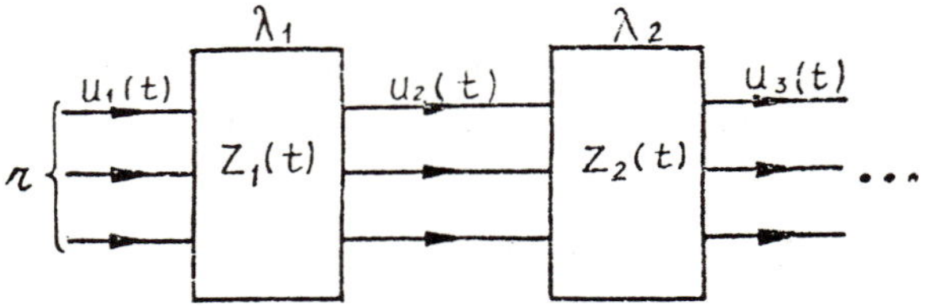

Fig. 13

It should be emphasized that relation (8.19) is the spectral resolution of a process into interconnected elementary open oscillators of the form (8.28) with complex frequencies λ_k. This expansion (8.19) differs from the well-known Loève-Karhunen expansion, which does not have a direct physical meaning [9, 21, 26].

§ 2. SPECTRAL RESOLUTIONS OF STOCHASTIC PROCESSES OF CLASS C_r

Let $z(t) = e^{itA}z_0$ be a linearly representable dissipative process with a completely non-self-adjoint operator A of finite non-Hermitian rank r whose spectrum consists of a single point at the origin. Such processes will be said to be of *class* C_r.

We include A in a complex $X = (A, H, g_1, \ldots, g_r, I)$, where $g_\alpha = \sqrt{\omega_\alpha} a_\alpha$ $(\alpha = 1, 2, \ldots, r)$, $(a_\alpha, a_\beta) = \delta_{\alpha\beta}$. We cite without proof the following theorem, which sharpens Theorem 5.6 and is needed for the sequel.

Theorem 8.3. *A complex X having the properties indicated above is unitarily equivalent to the complex*

$$\dot{X} = (\dot{A}, L_2(0, l), \dot{g}_1(x), \ldots, \dot{g}_r(x), I), \tag{8.30}$$

where

$$\dot{A}f(x) = i\int_0^x f(\xi)\, q(\xi)\, q^*(x)\, d\xi \quad (0 \leqslant x \leqslant l,\ f(x) \in L_2(0, l)), \tag{8.31}$$

in which $q(x) = \| \bar{\varphi}_1(x), \ldots, \bar{\varphi}_r(x) \|$ $(\varphi_\alpha(x) \in L_2(0, l))$ *is a row matrix such that*

$$\int_0^l q^*(x)\, q(x)\, dx = \| \omega_\alpha \delta_{\alpha\beta} \| \tag{8.32}$$

and

$$\sum_{\alpha=1}^r |\varphi_\alpha(x)|^2 \equiv 1 \quad (0 \leqslant x \leqslant l). \tag{8.33}$$

The channel elements $\dot{g}_\alpha(x)$ *have the form*

$$\dot{g}_\alpha(x) = \varphi_\alpha(x) \quad (0 \leqslant x \leqslant l). \tag{8.34}$$

The fact that model (8.31) has a variable upper limit of integration as opposed to a variable lower limit in model (5.75) is obviously of no significance, since it is possible to pass from one model to the other by substituting $l-x$ for x $(0 \leqslant x \leqslant l)$.

We introduce the function

$$\dot{Z}_{\Delta}(x) = \begin{cases} 1, & x' \leqslant x \leqslant x'', \\ 0 & x \,\overline{\in}\, [x', x''], \end{cases} \tag{8.35}$$

where $\Delta = [x', x'']$ $(0 \leqslant x' < x'' \leqslant l)$.

Clearly,

$$(\dot{Z}_{\Delta_1}, \dot{Z}_{\Delta_2}) = d(\Delta_1 \cap \Delta_2), \tag{8.36}$$

where d is the length of the interval $\Delta_1 \cap \Delta_2$.

It follows that $\dot{g}_{\alpha}$ and $f(x)$, as elements of the Hilert space $L_2(0, l)$, can be written in the form

$$\dot{g}_{\alpha} = \int_0^l \varphi_{\alpha}(x)\, d\dot{Z}_{[0,\, x]}, \tag{8.37}$$

$$f = \int_0^l f(x)\, d\dot{Z}_{[0,\, x]}, \tag{8.38}$$

in which

$$f(x) = \frac{d}{dx}(f, \dot{Z}_{[0,\, x]}). \tag{8.39}$$

Equations (6.38) for the open system associated with the complex (8.30) have the form

$$i\frac{\partial f}{\partial t} + i\int_0^x f(\xi, t) \sum_{\alpha=1}^{r} \overline{\varphi_{\alpha}(\xi)}\, \varphi_{\alpha}(x)\, d\xi = \sum_{\alpha=1}^{r} u_{\alpha}(t)\, \varphi_{\alpha}(x), \tag{8.40}$$

$$f|_{t=0} = f_0(x), \tag{8.41}$$

$$v_\alpha(t) = u_\alpha(t) - i\int_0^l f(\xi, t)\overline{\varphi_\alpha(\xi)}\,d\xi. \tag{8.42}$$

From equations (8.40) and (8.42) we can pass to the equivalent system of partial differential equations

$$i\frac{\partial f}{\partial t} = \sum_{\alpha=1}^{r} u_\alpha(x, t)\,\varphi_\alpha(x), \tag{8.43}$$

$$i\frac{\partial u_\alpha}{\partial t} = \overline{\varphi_\alpha(x)}\,f, \tag{8.44}$$

where

$$u_\alpha(x, t) = u_\alpha(0, t) - i\int_0^x f(\xi, t)\overline{\varphi_\alpha(\xi)}\,d\xi, \tag{8.45}$$

with

$$u_\alpha(0, t) \equiv u_\alpha(t), \quad u_\alpha(l, t) = v_\alpha(t). \tag{8.46}$$

Since $A = U\dot{A}U^{-1}$, where U is a unitary operator mapping $L_2(0, l)$ onto H, we obtain upon letting

$$Z_x = U\dot{Z}_{[0, x]}, \tag{8.47}$$

the following theorem.

Theorem 8.4. *For every stochastic process $z(t)$ of class C_r there exist a stochastic spectral measure Z_x $(0 \leqslant x \leqslant l)$ and a set of r functions $\varphi_\alpha(x)$ $(\alpha = 1, 2, \ldots r)$ satisfying the following conditions:*

1) $$M(\Delta_1 Z\overline{\Delta_2 Z}) = \rho(\Delta_1 \cap \Delta_2), \tag{8.48}$$

where $\Delta_k Z$ $(k = 1, 2)$ is the increase in Z_x over the interval Δ_k and $\rho(\Delta_1 \cap \Delta_2)$ is the length of the common part of the intervals Δ_k;

2) $$\sum_{\alpha=1}^{r} |\varphi_\alpha(x)|^2 \equiv 1 \quad (0 \leqslant x \leqslant l); \tag{8.49}$$

3) $$\int_0^l \varphi_\alpha(x)\overline{\varphi_\beta(x)}\,dx = \omega_\alpha\delta_{\alpha\beta}, \tag{8.50}$$

and such that the process $z(t)$ can be represented in the form

$$z(t) = \int_0^l f(x, t)\, dZ_x, \tag{8.51}$$

where the function $f(x, t)$ *is determined from the system of equations*

$$i \frac{\partial f}{\partial t} = \sum_{\alpha=1}^{r} u_\alpha(x, t)\, \varphi_\alpha(x), \tag{8.52}$$

$$i \frac{\partial u_\alpha}{\partial t} = \overline{\varphi_\alpha(x)}\, f(x, t) \tag{8.53}$$

and the conditions

$$f(x, 0) = f_0(x) \quad (f_0(x) \in L_2(0, l)), \tag{8.54}$$

$$u_\alpha(0, t) = 0 \quad (\alpha = 1, 2, \ldots, r). \tag{8.55}$$

We return now to the system of equations (8.52) and (8.53). Eliminating $f(x, t)$, we arrive at the following system of equations of hyperbolic type for the $u_\alpha(x, t)$:

$$\frac{\partial^2 u_\alpha}{\partial x\, \partial t} + \sum_{\beta=1}^{r} \overline{\varphi_\alpha(x)}\, \varphi_\beta(x)\, u_\beta = 0, \tag{8.56}$$

with the conditions

$$u_\alpha(0, t) = u_\alpha(t), \tag{8.57}$$

$$u_\alpha(x, 0) = u_\alpha(0) - i \int_0^l f_0(\xi)\, \overline{\varphi_\alpha(\xi)}\, d\xi. \tag{8.58}$$

Since, according to (7.10), the i.c.f. of $z(t)$ has the form

$$w(t, s) = \sum_{\alpha=1}^{r} \Phi_\alpha(t)\, \overline{\Phi_\alpha(s)}, \tag{8.59}$$

where

$$\Phi_\alpha(t) = (e^{itA} z_0, g_\alpha), \tag{8.60}$$

by setting $u_\alpha(0, t) \equiv 0$ in (8.57), we obtain the following expressions for $\Phi_\alpha(t)$:

$$\Phi_\alpha(t) = iu_\alpha(l, t), \tag{8.61}$$

which implies

$$w(t, s) = \sum_{\alpha=1}^{r} u_\alpha(l, t)\overline{u_\alpha(l, s)}. \tag{8.62}$$

With the use of these relations we can formulate a corollary of Theorem 8.4.

Corollary. To each process $z(t) \in C_r$ there corresponds a system of equations of the form

$$\frac{\partial^2 u_\alpha}{\partial x\, \partial t} + \sum_{\beta=1}^{r} \overline{\varphi_\alpha(x)}\,\varphi_\beta(x)\,u_\beta = 0 \tag{8.63}$$

with the conditions

$$u_\alpha(x, 0) = -i\int_0^x f_0(\xi)\,\overline{\varphi_\alpha(\xi)}\,d\xi, \tag{8.64}$$

$$u_\alpha(0, t) = 0, \tag{8.65}$$

the solution of which permits the determination of the function $f(x, t)$ *in the spectral resolution* (8.51) *by means of the formula*

$$f(x, t) = i\sum_{\alpha=1}^{r} \varphi_\alpha(x)\frac{\partial u_\alpha}{\partial x} \quad (0 \leqslant x \leqslant l, \quad 0 \leqslant t < \infty). \tag{8.66}$$

The i.c.f. of $z(t)$ *is determined in this connection by the equality*

$$w(t, s) = \sum_{\alpha=1}^{r} u_\alpha(l, t)\overline{u_\alpha(l, s)}. \tag{8.67}$$

We note that the spectral resolution (8.51) can be regarded as the continuous analog of the spectral representation (8.19) of complete processes, the system of difference equations (8.21) and (8.23) being replaced in the present case by a system of partial differential equations.

Suppose now $z(t)$ is a process of class K_r, i.e. a linearly representable process $z(t) = e^{itA}z_0$ in which A is the internal operator of a complex $X = X_1 \vee \vee X_2$ $(X_1 \in C_r(l),\ X_2 \in D_r(\Lambda))$ (Chapter V, §5). Since spectral resolutions of the form (8.19), (8.51) have already been obtained for the systems F_{X_1} and F_{X_2}, by making use of the fact that to a product $X_1 \vee X_2 = X$ of complexes there corresponds a coupling $F_{X_1} \vee F_{X_2} = F_X$ of open systems, we can obtain the following theorem.

Theorem 8.5. *For every stochastic process $z(t)$ of class K_r there exist a stochastic spectral measure Z_x $(0 \leqslant x \leqslant l)$ and a sequence of uncorrelated random variables Z_k $(k = 1, 2, \dots)$ such that*

1) $$M(Z_x \overline{Z}_k) = 0 \quad (0 \leqslant x \leqslant l,\ k = 1, 2, \dots); \tag{8.68}$$

2) $$Z(t) = \int_0^l f(x, t)\, dZ_x + \sum_{k=1}^{\infty} \psi_k(t) Z_k, \tag{8.69}$$

where $f(x, t)$ is determined from the system of equations (8.52) *and* (8.53) *and conditions* (8.54) *and* (8.55) *while $\psi_k(t)$ is found from the system of difference equations* (8.21) *and* (8.23), *for which the condition $u_{1,\alpha}(l) \equiv 0$ should be replaced by the condition*

$$u_{1,\alpha}(t) = u_\alpha(l, t), \tag{8.70}$$

where $u_\alpha(x, t)$ is the solution of the system of equations (8.63) *and the condition $u_\alpha(0, t) \equiv 0$.*

We note that analogous spectral resolutions of a somewhat more complicated form can be obtained for linearly representable processes with an arbitrary (nondissipative) linear operator A satisfying the condition dim (Im $A)H < \infty$ as well as for the wider class of linearly representable processes in which the operator A has a nuclear imaginary component.

NOTES TO CHAPTERS VII–VIII

The results of these chapters are due to M.S. Livshits and A.A. Yantsevich (§3 of Chapter VII is due to S.G. Mil'gram). Theorems 7.6–7.9 have been extended by

K.P. Kirčev [22] to linearly representable dissipative processes with an operator A that is unitarily equivalent to a triangular model of form (5.71). In this case an important role is played by the asymptotic behavior of the process as $t \to \infty$. Fundamental results were obtained in this direction by L.A. Sahnovič [37].

Theorem 8.3 was not formulated directly in the book [4], but it follows from Theorems 24,3 and 34,1 of the same monograph.

It is of interest to extend the obtained results to the case of an unbounded operator A by using the theory of triangular models of unbounded operators [24[1, 2, 8]].

We consider in particular an important class of such operators. Let A be an unbounded closed simple dissipative operator without a spectrum, with a dense domain of definition D_A and with $\dim \Delta_A = \Gamma < \infty$.

Consider A^{-1}. This operator exists, has a spectrum consisting of a single point at the origin and is dissipative and simple. By Theorem 5.12 it is unitarily equivalent to an operator induced on some invariant subspace by the operator

$$\dot{B} = -i\int_0^x \ldots d\xi \ldots \oplus \ldots \oplus -i\int_0^x \ldots d\xi$$

in the space $\dot{L}_2 = L_2^{(1)}(0, l) \oplus \ldots \oplus L_2^{(r)}(0, l)$.

The inverse of $\dot{B}$ clearly has the form

$$\dot{B}^{-1} = \tilde{D} = i\frac{d}{dx} \oplus \ldots \oplus i\frac{d}{dx},$$

while its domain of definition consists of functions that are absolutely continuous in each finite part of the interval $(0, l)$ and which together with their first derivatives belong to $L_2(0, l)$.

Theorem. *In order for a stochastic process $z(t)$ to be linearly representable with an unbounded dissipative operator without a spectrum and of rank r it is necessary and sufficient that*

$$w(t, s) = \sum_{\alpha=1}^{r} \Phi_\alpha(t)\overline{\Phi_\alpha(s)}.$$

where $\Phi_\alpha(t)$ is any function with compact support belonging to $L_2(0, l)$ $(\Phi_\alpha(t) \equiv 0,\ t \geqslant l)$.

An example of such a process is a stochastic process satisfying a linear stochastic differential equation with constant coefficients

$$\sum_{k=1}^{n} a_k \frac{d^k z}{dt^k} = \xi(t),$$

where $\xi(t)$ is white noise, and the conditions

$$z(l) = z'(l) = \ldots = z^{(n-1)}(l) = 0, \quad z(t) \equiv 0, \quad t > l.$$

CHAPTERS IX–X

The results presented in these chapters are due to V. K. Dubovoĭ [13[1,2]]. We note that [13[2]] contains a study of Weyl families of operator colligations that are invariant with respect to a reversal of the time and a reflection of the spatial coordinates.

CHAPTER IX

INVARIANT OPERATOR COLLIGATIONS

§1. AUXILIARY CONCEPTS

1. All of the spaces considered below are assumed to be finite-dimensional linear spaces over the field of complex numbers.

We recall that a space is said to be *unitary* if a positive definite metric is defined on it. One says that an *indefinite metric* is defined on a space H if there is associated with each pair of elements $f,\ h \in H$ a complex number $(f,\ h)$ which satisfies the following requirements:

a) $(f,\ h) = \overline{(h,\ f)}$;
b) $(\alpha_1 f_1 + \alpha_2 f_2, h) = \alpha_1 (f_1, h) + \alpha_2 (f_2, h)$;
c) if $(f, h) = 0$ for a fixed $f \in H$ and all $h \in H$ then $f = 0$;
d) there exist elements $f_1,\ f_2 \in H$ such that $(f_1,\ f_1) > 0,\ (f_2,\ f_2) < 0$.

A space on which a definite or indefinite metric is defined will be called a *psuedounitary space*. The adjoint of a linear operator A acting in a pseudo-unitary space will be denoted by A^+.

In the sequel a colligation (A, H, φ, E, μ) will be written more briefly in the form (A, H, φ, E) under the assumption that E is a pseudounitary space with a definite or indefinite metric (u, v) $(u, v \in E)$. It is also assumed in contrast to previous assumptions that the metric in H can be indefinite.

Let H and E be pseudounitary spaces and consider a pair of linear operators $A : H \to H$ and $\varphi : H \to E$. We recall that an aggregate (A, H, φ, E) is an *operator colligation* if

$$\frac{1}{i}(A - A^+) = \varphi^+\varphi \tag{9.1}$$

where φ^+ is the adjoint of φ and acts from E into H.

The spaces H, E and $\Delta_{\varphi+} = \varphi^+E$ will be called the *internal, external* and *channel spaces* respectively, while the operators A and φ will be called the *internal* and *channel operators.*

In a number of problems connected with open systems there exists a group with linear representations acting in the spaces H and E that leave the corresponding open system invariant. The condition that the open system be invariant reduces to the notion of an invariant operator colligation. In the present chapter the notion of an invariant operator colligation is introduced, criteria for the invariance of operator colligations are given, the possibility of an invariant operator colligation decomposing into invariant operator colligations is investigated and indecomposable invariant operator colligations are described for the case of normal representations.

The *characteristic operator function* (c.o.f.) of a given operator colligation $X = (A, H, \varphi, E)$ is

$$S_X(\lambda) = I - i\varphi(A - \lambda I)^{-1}\varphi^+ .$$

It is clearly defined and holomorphic on the set G_A of regular points of the operator A, and its values are linear operators acting in the space E.

We recall that a colligation (A, H, φ, E) is said to be *simple* if the linear span of vectors of the form $A^n\varphi^+v$ $(n = 0,1 \ldots$ and $v \in E)$ coincides with all of H.

We use the following notation: D_T is the domain of definition of an operator T, Δ_T is the range of an operator T and $\operatorname{Ker} T$ is the set of vectors $f \in D_T$ for which $Tf = 0$.

2. Suppose H is a pseudounitary space and $\tilde{H}$ is a subspace of H. If property c) is satisfied for $f, h \in \tilde{H}$, we will say that $\tilde{H}$ is *nondegenerate*. In this case the metric defined in H clearly induces a metric in $\tilde{H}$.

Further, the *annihilator* of a subspace $L \subset H$ is defined as the set L^0 of vectors $h \in H$ such that $(f, h) = 0$ for all $f \in L$. It is not difficult to see that L^0 is a subspace of H and that $L \subset (L^0)^0$. We have

Theorem 9.1. *Suppose H is a pseudounitary space and L is a subspace of H. Then*

1) $\dim L + \dim L^0 = \dim H$,
2) $(L^0)^0 = L$.

It can happen that $L \cap L^0 \neq 0$, but only if L is degenerate. Otherwise $L \cap L^0 = 0$ and Theorem 9.1 implies $H = L \dotplus L^0$. In this case L^0 is called the orthogonal complement of L and one writes $H = L \oplus L^0$.

Corollary 1. *Suppose B is an operator in a pseudounitary space H. Then Δ_{B^+} is the annihilator of* $\operatorname{Ker} B$.

Proof. *Suppose M is* the annihilator of $\operatorname{Ker} B$, $f \in \operatorname{Ker} B$, $h \in \Delta_{B^+}$ and $h = B^+h_1$. Then $(f, h) = (f, B^+h_1) = (Bf, h_1) = 0$. Hence $\Delta_{B^+} \subset M$. Since

$$\dim \Delta_{B^+} = \dim H - \dim \operatorname{Ker} B^+ = \dim H - \dim \operatorname{Ker} B,$$

by taking into account Theorem 9.1, we get $\Delta_{B^+} = M$.

Corollary 2. $\operatorname{Ker} B$ *is the annihilator of* Δ_{B^+}.

Suppose A is an operator acting in a pseudounitary space H. We will say that a nondegenerate subspace $\hat{H} \subset H$ reduces A, if both $\hat{H}$ and $H \ominus \hat{H}$ (i.e. the orthogonal complement of $\hat{H}$) are invariant under A.

An operator U acting in a space H is said to be *unitary* if $(Uf, Uh) = (f, h)$. for all $f, h \in H$ and $\Delta_U = H$.

3. A representation $g \to T_g$ of an arbitrary group G on a unitary space is said to be *normal* if the orthogonal complement of every subspace that is invariant under the operators T_g is also invariant under these operators. The following assertion, whose proof is left to the reader, describes the structure of a normal representation.

Lemma 9.1. *A representation* $g \to T_g$ *on a unitary space is normal if and only if*

1) *it splits into an orthogonal sum of irreducible representations* $g \to T_g^{(i)}$ $(i = 1, 2, \ldots, n)$:

$$T_g = T_g^{(1)} \oplus T_g^{(2)} \oplus \ldots\ldots \oplus T_g^{(n)}, \tag{9.2}$$

2) *equivalent representations in the sum* (9.2) *are unitarily equivalent.*

Lemma 9.1 and the well-known Schur's lemma imply

Lemma 9.2. *If* H *is a unitary space and an operator* C *acting in* H *commutes with the operators* T_g *of a normal representation* $g \to T_g$, *i.e.* $T_g C = C T_g$, *then* $T_g C^* = C^* T_g$.

Only completely reducible representations are considered in the sequel. A normal representation, as is easily seen, is completely reducible.

§2. INVARIANCE CRITERIA FOR OPERATOR COLLIGATIONS

Let G be an arbitrary group. A space H on which a representation $U: g \to U_g$, is given will be regarded as a *module* over G and denoted by H_U. We will also say that the module H_U *is induced* by the representation $g \to U_g$. A space G on which a pair of representations $\tilde{U} : g \to \tilde{U}_g$, $U : g \to U_g$, $g \in G$, is given will be called a *bimodule* over H and denoted by $H_{\tilde{U},U}$. In this case we will say that the bimodule $H_{\tilde{U},U}$ is induced by the representations $g \to \tilde{U}_g$ and $g \to U_g$. Thus $H_{\tilde{U}}$ and H_U are modules over G if $H_{\tilde{U},U}$ is a bimodule over G.

Let H'_U and $H''_{\tilde{U}}$ be two modules over G. We will say that an operator T acts from the module H'_U into the module $H''_{\tilde{U}}$ if $D_T = H'$, $\Delta_T \subset H''$ and

$$\tilde{U}_g T = T U_g \quad \text{for all } g \in G.$$

If $H' = H'' = H$, we will say that T acts in the bimodule $H_{\tilde{U},U}$. If, in addition, $\tilde{U} = U$, we will say that T acts in the module H_U.

Let H_U be a module over G. A subspace $\tilde{H} \subset H$ that is invariant with respect to the representation $g \to U_g$ will be called a *submodule* of H_U.

Definition. An operator colligation $(A, H_{\tilde{U},U}, \varphi, E_V)$, where $H_{\tilde{U},U}$ is a bimodule over G and E_V is a module over G, will be said to be *invariant* with respect to G if A acts in $H_{\tilde{U},U}$, φ acts from H_U into E_V and φ^+ acts from E_V into $H_{\tilde{U}}$.

Thus a colligation $(A, H_{\tilde{U},U}, \varphi, E_V)$ is invariant with respect to a group G if and only if the following relations are satisfied:

$$\tilde{U}_g A = A U_g, \tag{9.3}$$

$$V_g \varphi = \varphi U_g, \tag{9.4}$$

$$\tilde{U}_g \varphi^+ = \varphi^+ V_g. \tag{9.5}$$

The representations $g \to \tilde{U}_g$ and $g \to U_g$ of an invariant colligation $(A, H_{\tilde{U},U}, \varphi, E_V)$ will be called its *internal representations* while the representation $g \to V_g$ will be called its *external representation*. An invariant operator colligation will be said to be *strongly invariant* if $\tilde{U} = U$.

We have the following assertion.

Lemma 9.3. *Let H_U be a module over G and let φ be an operator acting from the space H into a space E. If $\operatorname{Ker}\varphi$ is a submodule of H_U, it is possible to define on $\tilde{E} = \Delta_\varphi$ a representation $g \to V_g$, $g \in G$, such that the operator φ will act from the module H_U into the module E_V; the representation $g \to V_g$ in this connection is uniquely determined by the module H_U and the operator φ.*

Proof. There exists a complement of $\operatorname{Ker}\varphi$ that is also a submodule of H_U. We denote it by H^c and let φ_1 denote the restriction of φ to H^c. We define the representation $g \to V_g$ on Δ_φ by putting $V_g f = \varphi_1 U_g \varphi_1^{-1} f$, $f \in \Delta_\varphi$. Rewriting this relation in the form $V_g \varphi = \varphi U_g$, we convince ourselves of the validity of the lemma.

Theorem 9.2. *If $(A, H_{\tilde{U},U}, \varphi, E_V)$ is an operator colligation that is invariant with respect to a group G then*

$$A \text{ and } A^+ \text{ act in the bimodule } H_{\tilde{U},U}, \tag{9.6}$$

$$\text{Ker } \varphi \textit{ is a submodule of } H_U, \tag{9.7}$$

$$\Delta_{\varphi^+} \textit{ is a submodule of } H_{\tilde{U}}. \tag{9.8}$$

Suppose $(A, H_{\tilde{U},U}, \varphi, E)$ *is an operator colligation,* $H_{\tilde{U},U}$ *is a bimodule over* G *and conditions* (9.6)-(9.8) *are satisfied. Then a representation* $g \to V_g$, $g \in G$, *such that the colligation* $(A, H_{\tilde{U},U}, \varphi, E_V)$ *will be invariant with respect to the group* G *can be defined on* E.

Proof. If $(A, H_{\tilde{U},U}, \varphi, E_V)$ is a colligation that is invariant with respect to a group G, conditions (9.7) and (9.8) follow from (9.4) and (9.5) respectively. Further, (9.4) implies $\varphi^+ V_g \varphi = \varphi^+ \varphi V_g$, which together with (9.5) implies

$$\tilde{U}_g \varphi^+ \varphi = \varphi^+ \varphi U_g. \tag{9.9}$$

Taking into account the colligation condition (9.1), we rewrite (9.9) in the form

$$\tilde{U}_g (A - A^+) = (A - A^+) U_g. \tag{9.10}$$

Condition (9.6) follows from (9.10) and (9.3). Thus the first part of the theorem is proved.

Suppose now $(A, H_{\tilde{U},U}, \varphi, E)$ is an operator colligation satisfying conditions (9.6)-(9.8). From (9.7) and Lemma 9.3 it follows that a representation $g \to V_g^{(1)}$, $g \in G$, satisfying the condition

$$V_g^{(1)} \varphi = \varphi U_g \tag{9.11}$$

can be defined on Δ_φ. Hence, noting that (9.6) implies the relation

$$\tilde{U}_g \varphi^+ \varphi = \varphi^+ \varphi U_g, \tag{9.12}$$

we get $\tilde{U}_g \varphi^+ \varphi = \varphi^+ V_g^{(1)} \varphi$.

Consequently,

$$\tilde{U}_g \varphi^+ f = \varphi^+ V_g^{(1)} f, \; f \in \Delta_\varphi. \tag{9.13}$$

Relation (9.12) implies the invariance of $\Delta_{\varphi^+\varphi}$ under the $\tilde{U}_g$. Since $\Delta_{\varphi^+\varphi} \subset \Delta_{\varphi^+}$ and Δ_{φ^+} is invariant under the U_g, there exists a subspace H^c that is invariant under the $\tilde{U}_g$ and such that $\Delta_{\varphi^+} = \Delta_{\varphi^+\varphi} \dot{+} H^c$. Consider the complete preimage $(\varphi^+)^{-1}H^c = E^c$ of H^c. Let E_0^c be one of the complements of Ker φ^+ in E^c. Clearly, $E_0^c \cap \Delta_\varphi = 0$. Consequently, E can be represented in the form

$$E = \Delta_\rho \dot{+} E_0^c \dot{+} \tilde{E}, \tag{9.14}$$

where $\tilde{E} \subset \text{Ker}\, \varphi^+$.

The decomposition (9.14) determines a decomposition

$$I = P_1 + P_2 + P_3, \tag{9.15}$$

where P_1 is the projection along (i.e. with null space) $E_0^c \dot{+} \tilde{E}$ onto Δ_φ, P_2 is the projection along $\Delta_\varphi \dot{+} \tilde{E}$ onto E_0^c and P_3 is the projection along $\Delta_\varphi \dot{+} E_0^c$ onto $\tilde{E}$.

Let φ_1^+ be the restriction of φ^+ to E_0^c. By virtue of the arguments carried out above we can consider a representation $g \to V_g^{(2)}$, $g \in G$, on E_0^c by putting $V_g^{(2)}f = (\varphi_1^+)^{-1}\tilde{U}_g\varphi_1^+ f$, $f \in E_0^c$. From this we get

$$\tilde{U}_g\varphi^+P_2 = \varphi^+V_g^{(2)}P_2. \tag{9.16}$$

We now define a representation $g \to V_g$ on E by putting $V_g = V_g^{(1)}P_1 + V_g^{(2)}P_2 + V_g^{(3)}P_3$, where $g \to V_g^{(3)}$ is any representation of the group G on $\tilde{E}$ (for example, we can put $V_g^{(3)} = I$ for all $g \in G$).

From equalities (9.15), (9.13), (9.16) and the fact that $\varphi^+P_3 = 0$, $V_g^{(3)}P_3 = P_3V_g$ we get $\tilde{U}_g\varphi^+ = \varphi^+V_g$. Since (9.11) can be rewritten in the form $V_g\varphi = \varphi U_g$, the theorem is completely proved.

Definition. An operator colligation $(A, H_{\tilde{U},U}, \varphi, E)$ satisfying conditions (9.6)-(9.8) will be said to be *preinvariant.*

Lemma 9.4. *Suppose* $X = (A, H_{\tilde{U},U}, \varphi, E)$ *is an operator colligation and the operators* A *and* A^+ *act in the bimodule* $H_{\tilde{U},U}$. *Then* X *is preinvariant if* Δ_φ *is a nondegenerate subspace of* E.

Proof. We have

$$\tilde{U}_g A = A U_g, \ \tilde{U}_g A^+ = A^+ U_g.$$

Hence, using the colligation condition (9.1), we get

$$\tilde{U}_g \varphi^+ \varphi = \varphi^+ \varphi U_g. \tag{9.17}$$

Since $\operatorname{Ker} \varphi = \operatorname{Ker} \varphi^+\varphi$, $\Delta_{\varphi^+} = \Delta_{\varphi^+\varphi}$ in the present case, conditions (9.7) and (9.8) follow from (9.17). The lemma is proved.

Lemma 9.5. *Suppose (A, H_U, φ, E_V) is an operator colligation. Then the linear span of vectors of the form*

$$A^n \varphi^+ f \ (n = 0, \ 1, \ \ldots \ ; \ f \in E) \tag{9.18}$$

coincides with the linear span of vectors of the form

$$A^{+n} \varphi^+ f \ (n = 0, \ 1, \ \ldots \ ; \ f \in E). \tag{9.19}$$

Proof. Let H_1 (H_2) be the linear span of vectors of the form (9.18) ((9.19)) and let H_1^0 be the annihilator of H_1. Since H_1 is clearly invariant under A, we conclude that H_1^0 is invariant under A^+. From the colligation condition (9.1) it follows that the range of the operator $A - A^+$ belongs to H_1. Hence $(A - A^+) h = 0$, $h \in H_1^0$. Thus $Ah = A^+h$, $h \in H_1^0$, so that H_1^0 is also invariant under A. This implies that H_1 is invariant under A^+. Consequently, $H_2 \subseteq H_1$. The fact that $H_1 \subseteq H_2$ is proved analogously. The lemma is proved.

Corollary. *If $X = (A, H, \varphi, E)$ is a simple operator colligation and $\varphi A^n h = 0$ $(n = 0, 1, \ldots)$ then $h = 0$.*

Proof. From the simplicity of the colligation X and Lemma 9.5 we find that any vector $\tilde{h} \in H$ can be represented in the form

$$\tilde{h} = \sum_p A^{+n_p} \varphi^+ v_p \quad (v_p \in E).$$

But then

$$(h, \tilde{h})_H = (h, \sum_p A^{+n_p}\varphi^+ v_p)_H = \sum_p (\varphi A^{n_p} h, v_p)_E = 0,$$

which implies that $h = 0$.

Theorem 9.3. *Suppose* $X = (A, H_U, \varphi, E_V)$ *is an operator colligation that is strongly invariant with respect to a group G and* $S_X(\lambda)$ *is the c.o.f. of* X. *Then*

$$V_g S_X(\lambda) = S_X(\lambda)\, V_g, \tag{9.20}$$

i.e. the operator function $S_X(\lambda)$ *acts in the module* E_V.

If $X = (A, H, \varphi, E_V)$ *is* a simple operator colligation, E_V is a module over G and condition (9.20) is satisfied, then a representation $g \to U_g$, $g \in G$, such that the colligation (A, H_U, φ, E_V) will be strongly invariant with respect to the group G can be defined on H. This representation is uniquely determined by the colligation (A, H, φ, E_V).

Proof. Relation (9.20) follows from the invariance conditions (9.3)-(9.5).

Suppose now $X = (A, H, \varphi, E_V)$ is a simple operator colligation whose c.o.f. satisfies relation (9.20). We will show that in this case the subspaces $\mathrm{Ker}\,(A^n\varphi^+)$ $(n = 0, 1 \ldots)$ are invariant with respect to the representation $g \to V_g$.

To this end we note that for sufficiently large $|\lambda|$

$$S_X(\lambda) = I + i\sum_{n=0}^{\infty} \frac{\varphi A^n \varphi^+}{\lambda^{n+1}}.$$

Therefore, taking into account (9.20), we have

$$V_g \varphi A^n \varphi^+ = \varphi A^n \varphi^+ V_g \; (n = 0, 1 \ldots). \tag{9.21}$$

We rewrite (9.21) in the form

$$V_g \varphi A^m A^n \varphi^+ = \varphi A^m A^n \varphi^+ V_g \; (n, m = 0, 1, \ldots). \tag{9.22}$$

By fixing n and varying $m = 0,1, \ldots$ in (9.22) and, in addition, taking into account the simplicity of the colligation X and the Corollary of Lemma 9.5, we obtain the desired result.

From Lemma 9.3 it follows that a representation $g \to U_g^{(p)}$, $g \in G$, such that

$$U_g^{(p)} A^p \varphi^+ = A^p \varphi^+ V_g \quad (p = 0,1, \ldots) \tag{9.23}$$

can be defined on $\Delta_p = \Delta_{A^p \varphi^+}$.

Let us prove the relation

$$U_g^{(p)} | \Delta_{pq} = U_g^{(q)} | \Delta_{pq}, \tag{9.24}$$

where $\Delta_{pq} = \Delta_p \cap \Delta_q$. We have

$$V_g \varphi A^m A^p \varphi^+ = \varphi A^m A^p \varphi^+ V_g = \varphi A^m U_g^{(p)} A^p \varphi^+ \quad (m = 0,1, \ldots)$$

or

$$V_g \varphi A^m h = \varphi A^m U_g^{(p)} h, \; h \in \Delta_p \, (m = 0,1, \ldots). \tag{9.25}$$

Putting $p = q$, we get

$$V_g \varphi A^m h = \varphi A^m U_g^{(q)} h, \; h \in \Delta_q \quad (m = 0,1, \ldots). \tag{9.26}$$

If $h \in \Delta_{pq}$, relations (9.25) and (9.26) imply

$$\varphi A^m U_g^{(p)} h = \varphi A^m U_g^{(q)} h \quad (m = 0,1, \ldots). \tag{9.27}$$

Taking into account the simplicity of X and the Corollary of Lemma 9.5, we conclude from (9.27) that

$$U_g^{(p)} h = U_g^{(q)} h, \quad h \in \Delta_{pq}. \tag{9.28}$$

Since X is simple and $\dim H < \infty$, there exists an l such that

$$H = \Delta_0 + \Delta_1 + \ldots + \Delta_l.$$

From (9.28) we see that a representation $g \to U_g$ can be defined on H by putting

$$U_g \mid \Delta_p = U_g^{(p)} \ (p = 0,1, \ldots , l).$$

We now rewrite (9.23) in the form

$$U_g A^p \varphi^+ = A^p \varphi^+ V_g \ (p = 0,1, \ldots). \tag{9.29}$$

This implies

$$U_g A A^p \varphi^+ = A A^p \varphi^+ V_g = A U_g A^p \varphi^+ \ (p = 0,1, \ldots), \tag{9.30}$$

$$V_g \varphi A^p \varphi^+ = \varphi A^p \varphi^+ V_g = \varphi U_g A^p \varphi^+ \ (p = 0,1, \ldots). \tag{9.31}$$

By virtue of the simplicity of X it follows from (9.30) and (9.31) that $U_g A = AU_g$, $V_g \varphi = \varphi U_g$. Finally, putting $p = 0$ in (9.29), we get $U_g \varphi^+ = \varphi^+ V_g$, and the theorem is completely proved.

Corollary. *Suppose $X = (A, H, \varphi, E_V)$ is a simple operator colligation and E_V is the module over G induced by a unitary representation $g \to V_g$. If*

$$V_g S_X(\lambda) = S_X(\lambda) V_g,$$

then a unitary representation $g \to U_g$, $g \in G$ can be defined on H, such that the colligation (A, H_U, φ, E_V) will be strongly invariant with respect to the group G. This representation is uniquely determined by the colligation (A, H, φ, E_V).

Proof. By Theorem 9.3 there exists a representation $g \to U_g$ of G on H which is uniquely determined by the colligation (A, H, φ, E_V), such that the colligation (A, H_U, φ, E_V) is strongly invariant with respect to G. We will show that this representation is a unitary representation in the case under consideration.

The invariance conditions (9.3)-(9.5) and Theorem 9.2 imply

$$V_g\varphi A^{+m}A^n\varphi^+ = \varphi U_g A^{+m}A^n\varphi^+ = \varphi A^{+m}A^n U_g\varphi^+ = \varphi A^{+m}A^n\varphi^+ V_g.$$
$$(m,\ n = 0, 1,\ \ldots). \tag{9.32}$$

Further, by virtue of the simplicity of X any vectors $h_1,\ h_2 \in H$ can be represented in the form

$$h_1 = \sum_{i=1}^{n_1} A^{p_i}\varphi^+ f_i^{(1)};\ h_2 = \sum_{j=1}^{n_2} A^{q_j}\varphi^+ f_j^{(2)};\ f_i^{(1)},\ f_j^{(2)} \in E.$$

From relations (9.29) and (9.32) and the fact that $g \to V_g$ is a unitary representation we get

$$(U_g h_1,\ U_g h_2)_H = \left(U_g \sum_{i=1}^{n_1} A^{p_i}\varphi^+ f_i^{(1)},\ U_g \sum_{j=1}^{n_2} A^{q_j}\varphi^+ f_j^{(2)}\right)_H =$$
$$\sum_{i=1}^{n_1}\sum_{j=1}^{n_2} (A^{p_i}\varphi^+ V_g f_i^{(1)},\ A^{q_j}\varphi^+ V_g f_j^{(2)})_H =$$
$$\sum_{i=1}^{n_1}\sum_{j=1}^{n_2} (V_g f_i^{(1)},\ \varphi A^{+p_i}A^{q_j}\varphi^+ V_g f_j^{(2)})_E =$$
$$\sum_{i=1}^{n_1}\sum_{j=1}^{n_2} (V_g f_i^{(1)},\ V_g\varphi A^{+p_i}A^{q_j}\varphi^+ f_j^{(2)})_E =$$
$$\sum_{i=1}^{n_1}\sum_{j=1}^{n_2} (f_i^{(1)},\ \varphi A^{+p_i}A^{q_j}\varphi^+ f_j^{(2)})_E = (h_1,\ h_2)_H,$$

Q.E.D.

§3. PRODUCT OF INVARIANT OPERATOR COLLIGATIONS

We consider colligations $X_i = (A_i, H^{(i)}_{\tilde{U}^{(i)},U^{(i)}}, \varphi_i, E_V)$ $(i = 1,2)$ that are invariant with respect to a group G and let P_1 and P_2 denote the orthogonal projections onto $H^{(1)}$ and $H^{(2)}$ in the space $H = H^{(1)} \oplus H^{(2)}$. Thus P_1 is the projection along $H^{(2)}$ onto $H^{(1)}$ while P_2 is the projection along $H^{(1)}$ onto $H^{(2)}$.

We introduce the operators

$$A = A_1 P_1 + A_2 P_2 + i\varphi_2^+ \varphi_1 P_1, \quad \varphi = \varphi_1 P_1 + \varphi_2 P_2,$$

$$\tilde{U}_g = \tilde{U}_g^{(1)} P_1 + \tilde{U}_g^{(2)} P_2, \quad U_g = U_g^{(1)} P_1 + U_g^{(2)} P_2, \qquad (9.33)$$

acting in H and from H onto E.

The spaces H, E and operators A, φ form a colligation. In fact, since

$$\varphi_1^+ \varphi_1 = \frac{1}{i}\left(A_1 - A_1^+\right); \quad \varphi_2^+ \varphi_2 = \frac{1}{i}\left(A_2 - A_2^+\right);$$
$$A^+ = A_1^+ P_1 + A_2^+ P_2 - i\varphi_1^+ \varphi_2 P_2; \quad \varphi^+ = \varphi_1^+ + \varphi_2^+,$$

we get

$$\frac{1}{i}(A - A^+) = \varphi_1^+ \varphi_1 P_1 + \varphi_2^+ \varphi_2 P_2 + \varphi_1^+ \varphi_2 P_2 + \varphi_2^+ \varphi_1 P_1 =$$
$$(\varphi_1^+ + \varphi_2^+)(\varphi_1 P_1 + \varphi_2 P_2) = \varphi^+ \varphi.$$

We let $H_{\tilde{U},U}$ denote the bimodule induced by the representations $g \to \tilde{U}_g$, $g \to U_g$ and prove that the colligation $(A, H_{\tilde{U},U}, \varphi, E_V)$ is invariant with respect to the group G. In fact, since the colligations are invariant,

$$\tilde{U}_g A = \tilde{U}_g^{(1)} A_1 P_1 + \tilde{U}_g^{(2)} A_2 P_2 + i\tilde{U}_g^{(2)} \varphi_2^+ \varphi_1 P_1 =$$
$$A_1 U_g^{(1)} P_1 + A_2 U_g^{(2)} P_2 + i\varphi_2^+ V_g \varphi_1 P_1 =$$
$$A_1 U_g^{(1)} P_1 + A_2 U_g^{(2)} P_2 + i\varphi_2^+ \varphi_1 U_g^{(1)} P_1 = A U_g,$$
$$V_g \varphi = V_g \varphi_1 P_1 + V_g \varphi_2 P_2 = \varphi_1 U_g^{(1)} P_1 + \varphi_2 U_g^{(2)} P_2 = \varphi U_g,$$
$$\tilde{U}_g \varphi^+ = \tilde{U}_g^{(1)} \varphi_1^+ + \tilde{U}_g^{(2)} \varphi_2^+ = \varphi_1^+ V_g + \varphi_2^+ V_g = \varphi^+ V_g.$$

Let us agree to call the invariant colligation $X = (A_1 P_1 + A_2 P_2 + i\varphi_2^+ \varphi_1 P_1,$ $H_{\tilde{U},U}, \varphi_1 P_1 + \varphi_2 P_2, E_V)$ the *product* of the invariant colligations X_1 and X_2 and to write $X = X_1 \vee X_2$.

A direct verification shows that the formula $(X_1 \vee X_2) \vee X_3 = X_1 \vee (X_2 \vee X_3)$ is valid.

Definition. Let $H_{\tilde{U},U}$ be a bimodule over G and let H be a pseudounitary space. A nondegenerate subspace $\tilde{H}$ that is nontrivial, i.e. different from the zero subspace and from all of H, is said to *decompose* the bimodule $H_{\tilde{U},U}$ if $\tilde{H}$ and $H \ominus \tilde{H}$ are submodules of the modules H_U and $H_{\tilde{U}}$.

We now introduce the notion of a projection of an invariant colligation. Let $X = (A, H_{\tilde{U},U}, \varphi, E_V)$ be a colligation that is invariant with respect to the group G, let H' be a subspace that decomposes the bimodule $H_{\tilde{U},U}$ and let P' be the projection along $H \ominus H'$ onto H'. We consider in H' the operators $A'h = P'Ah$, $\tilde{U}'_g h = \tilde{U}_g h$, $U'_g h = U_g h$; $h \in H'$. In addition, we construct the mapping $\varphi' = \varphi P'$ of H' into E.

Since

$$\left.\begin{aligned} &(A')^+ h = P'A^+h, \ \varphi' h = \varphi h, \\ &\varphi'^+\varphi' h = P'\varphi^+\varphi h = P'\frac{1}{i}(A - A^+)h = \frac{1}{i}(A' - A'^+)h, \end{aligned}\right\} h \in H',$$

the spaces H', E and operators A', φ' form an operator colligation. Let $H'_{\tilde{U}',U}$ denote the bimodule induced by the representations $g \to \tilde{U}'_g$ and $g \to U'_g$. It is easily seen that the operator colligation $X' = (A', H'_{\tilde{U}',U'}, \varphi', E_V)$ is invariant with respect to the group G.

We will call the colligation X' the *projection* of X onto H' and write $X' = \mathrm{pr}_{H'}X$.

If $X_1 = (A_1, H^{(1)}_{\tilde{U}^{(1)},U^{(1)}}, \varphi_1, E_V)$ and $X_2 = (A_2, H^{(2)}_{\tilde{U}^{(2)},U^{(2)}}, \varphi_2, E_V)$ are operator colligations that are invariant with respect to a group G and $X = X_1 \vee X_2 = (A, H_{\tilde{U},U}, \varphi, E_V)$ then, as formulas (9.33) show, X_1 and X_2 are the projections of X onto $H^{(1)}$ and $H^{(2)}$ respectively, with $H^{(2)}$ being invariant under A. Conversely, an invariant colligation $X = (A, H_{\tilde{U},U}, \varphi, E_V)$ is the product of its projections $X_1 = (A_1, H^{(1)}_{\tilde{U}^{(1)},U^{(1)}}, \varphi_1, E_V)$ and $X_2 = (A_2, H^{(2)}_{\tilde{U}^{(2)},U^{(2)}}, \varphi_2, E_V)$ onto a subspace $H^{(1)}$ and onto a subspace $H^2 = H \ominus H^{(1)}$ that is invariant under A and decomposes the bimodule $H_{\tilde{U},U}$.

Definition. An invariant colligation will be said to be *indecomposable* if it is not representable in the form of a product of invariant operator colligations.

Definition. Let H be a pseudounitary space and let $H_{\tilde{U},U}$ be a bimodule over G. An operator acting in the bimodule $H_{\tilde{U},U}$ will be said to be *indecomposable* if there do not exist subspaces that are invariant under A and decompose the bimodule $H_{\tilde{U},U}$.

The arguments carried out above directly imply

Lemma 9.6. *An invariant colligation* $(A,\ H_{\tilde{U},U},\ \varphi,\ E_V)$ *is indecomposable if and only if its internal operator is indecomposable in the bimodule* $H_{\tilde{U},U}$.

Remark. All of the results and definitions of this section can be carried over to strongly invariant operator colligations by simply putting $\tilde{U}=U$ in the appropriate places.

§ 4. INDECOMPOSABLE OPERATORS

Lemma 9.7. *Let* H *be a unitary space, let* H_U *be a module over* G *induced by a normal representation* $g\to U_g,\ g\in G$, *and let* A *be an indecomposable operator acting in* H_U. *Then* $A=\lambda I$ *and the representation* $g\to U_g$ *is irreducible.*

Proof. We have the formula

$$U_gA=AU_g. \tag{9.34}$$

Let N_λ denote the eigenspace of A corresponding to the eigenvalue λ. From (9.34) we see that N_λ is invariant under the U_g. Since the representation $g\to U_g$ is normal, N_λ reduces the U_g. Consequently, N_λ decomposes H_U. Since A is indecomposable, $N_\lambda=H$ and $A=\lambda I$. Moreover, from this result we see that the representation $g\to U_g$ is irreducible. The lemma is proved.

Theorem 9.4. *Let* $(A,\ H_U,\ \varphi,\ E_V)$ *is an indecomposable operator colligation that is strongly invariant with respect to the group* G. *Suppose* H *is a unitary space and* $g\to U_g,\ g\in G$, *is a normal representation. Then* $A=\lambda I$, *the representation* $g\to U_g$ *is irreducible and either* $\Delta_{\varphi^+}=0$ *or* $\Delta_{\varphi^+}=H$.

This assertion follows from Lemma 9.6, Theorem 9.2 and Lemma 9.7.

Lemma 9.8. *Let* H *be a unitary space and let* $H_{\tilde{U},U}$ *be a bimodule over* G *induced by a pair of normal representations* $g\to\tilde{U}_g,\ \ g\to U_g,\ \ g\in G$. *Let* A *and* A^* *be operators acting in* $H_{\tilde{U},U}$ *and suppose* A *is singular and indecomposable. Then* $A=0$ *and the representations* $g\to\tilde{U}_g,\ \ g\to U_g$ *do not have a common nontrivial invariant subspace.*

Proof. We have the formulas

$$\tilde{U}_gA=AU_g, \tag{9.35}$$

$$\tilde{U}_gA^*=A^*U_g. \tag{9.36}$$

From (9.35) we see that Ker A is invariant under the U_g. From (9.36) it follows that Δ_{A^*} is invariant under the $\tilde{U}_g$. Since Ker $A \perp \Delta_{A^*}$ and the representations $g \to \tilde{U}_g$ and $g \to U_g$ are normal, Ker A decomposes $H_{\tilde{U},U}$. Further, Ker $A \neq 0$. Therefore Ker $A = H$ and $A = 0$. In addition, it is easily seen that the representations $g \to \tilde{U}_g$ and $g \to U_g$ do not have a common nontrivial invariant subspace. The lemma is proved.

Theorem 9.5. *Let* $(A, H_{\tilde{U},U}, \varphi, E_V)$ *be an indecomposable operator colligation that is invariant with respect to the group* G *and whose internal operator is singular. Suppose* H *is a unitary space and the representations* $g \to \tilde{U}_g$ *and* $g \to U_g$, $g \in G$, *are normal. Then* $A = 0$, *the representations* $g \to \tilde{U}_g$ *and* $g \to U_g$ *do not have a common nontrivial invariant subspace and either* $\Delta_{\varphi^+} = 0$ *or* $\Delta_{\varphi^+} = H$.

This assertion follows from Lemma 9.6, Theorem 9.2 and Lemma 9.8.

Let H be a unitary space and let $H_{\tilde{U},U}$ be a bimodule over G induced by normal representations. Suppose now that an indecomposable operator A acting in $H_{\tilde{U},U}$ is nonsingular. Then it follows from (9.35) that the representations $g \to U_g$ and $g \to \tilde{U}_g$ are equivalent. Suppose that A^* also acts in $H_{\tilde{U},U}$. We will describe the operator A for this case, assuming in addition that the representations $g \to \tilde{U}_g$ and $g \to U_g$ are unitarily equivalent, i.e.

$$\tilde{U}_g = UU_gU^{-1}, \quad U^* = U^{-1}, \tag{9.37}$$

From (9.35) and (9.36), taking into account (9.37), we get

$$U_gU^{-1}A = U^{-1}AU_g, \tag{9.38}$$

$$U_gU^{-1}A^* = U^{-1}A^*U_g. \tag{9.39}$$

Using Lemma 9.2, we rewrite (9.36) in the form

$$U_gAU = AUU_g. \tag{9.40}$$

Consequently,

$$U_gA^2 = U_gAUU^{-1}A = AUU_gU^{-1}A = AUU^{-1}AU_g = A^2U_g. \tag{9.41}$$

Let λ be an eigenvalue of A. Then λ^2 will be an eigenvalue of A^2. Consider the eigenspace N_{λ^2} of A^2 corresponding to the eigenvalue λ^2. From (9.41) we see that N_{λ^2} is invariant under the U_g. Since $\tilde{U}_g = AU_gA^{-1}$ it follows that N_{λ^2} is also invariant under the $\tilde{U}_g$. Consequently, N_{λ^2} decomposes $H_{\tilde{U},U}$. Since A is indecomposable, $N_{\lambda^2} = H$ and $A^2 = \lambda^2 I$, i.e. $A = \lambda B$, $B^2 = I$.

From the formula $\tilde{U}_g = AU_gA^{-1} = BU_gB^{-1}$ we see that the decomposability of A depends on the operator B and the representation $g \to U_g$ having a common nontrivial invariant subspace. The following example shows that such a subspace does not always exist.

Example. Consider a set of pairs $\{f;\ h\}$ in which f and h range over a unitary space H. We will regard these pairs as elements of the orthogonal sum $\tilde{H} = H_1 \oplus H_2$, where $H_1 = H_2 = H$.

We have $\tilde{H} = M_1 \oplus M_2$, where $M_1 = \{f,\ f\}$ and $M_2 = \{f,\ -f\}$, $f \in H$. Suppose given a representation $g \to U_g$ of a group G on $\tilde{H}$ such that

1) it is reducible and normal on M_1,

2) it is irreducible on M_2.

This representation is clearly normal on $\tilde{H}$.

We define an operator B by setting

$$Bf = f,\ f \in H_1;\ Bh = -h,\ h \in H_2.$$

Clearly, $BM_1 = M_2$, $BM_2 = M_1$ and $B = B^* = B^{-1}$.

Let $\tilde{U}_g = BU_gB^{-1}$. Then

$$\tilde{U}_gB = BU_g,\ \tilde{U}_gB^* = B^*U_g$$

and the representations $g \to \tilde{U}_g$ and $g \to U_g$ are normal and unitarily equivalent. It is not difficult to see that the operator B is indecomposable in the module $\tilde{H}_{\tilde{U},U}$.

We have thus proved

Theorem 9.6. *Let H be a unitary space and let $H_{\tilde{U},U}$ be a bimodule over G induced by a pair of normal and unitarily equivalent representations $g \to \tilde{U}_g$ and $g \to U_g$. Let A and A^* be operators acting in $H_{\tilde{U},U}$ and suppose A is nonsingular and indecomposable. Then $A = \lambda B$, where the operator B is such that*

$B^2 = I$ and it and the representation $g \to U_g$ do not have a common nontrivial invariant subspace.

We will prove the following assertion.

Theorem 9.7. *Under the conditions of the preceding theorem either $B = B^* = B^{-1}$ or $B^* = QB$, where the matric of the operator Q relative to some orthonormal basis has the form*

$$\begin{pmatrix} \alpha I & 0 \\ 0 & \frac{1}{\alpha} I \end{pmatrix}, \quad \alpha > 0, \ \alpha \neq 1.$$

Here $\dim M_\alpha = \dim M_{\frac{1}{\alpha}}$, where M_α and $M_{\frac{1}{\alpha}}$ are the eigenspaces of Q corresponding to the eigenvalues α and $\frac{1}{\alpha}$.

Proof. We have

$$\tilde{U}_g B = B U_g, \quad B^2 = I,$$
$$\tilde{U}_g = V U_g V^{-1}, \quad V^* = V^{-1}.$$

Let $C = V^{-1}B$. Clearly,

$$U_g C = C U_g. \tag{9.42}$$

Using Lemma 9.2, we rewrite (9.42) in the form

$$U_g C^* = C^* U_g. \tag{9.43}$$

From (9.42) and (9.43) we get

$$U_g C^* C = C^* C U_g. \tag{9.44}$$

Let $Q = B^*B = C^*C$. Then $B^* = QB$ and (9.44) implies

$$U_g Q = Q U_g. \tag{9.45}$$

If $B^* \neq B$ then $Q \neq I$. In this case the operator Q has at least one eigenvalue α such that $\alpha > 0$ and $\alpha \neq 1$. Let M_α be the corresponding eigenspace.

From the equalities $B^* = QB = B^{*-1} = BQ^{-1}$ we get $QBQ = B$. Consequently, $QBh = \frac{1}{\alpha} Bh$, $h \in M_\alpha$. Hence $BM_\alpha \subset M_{\frac{1}{\alpha}}$. Since these same arguments can be repeated in regard to $M_{\frac{1}{\alpha}}$, we get

$$BM_\alpha = M_{\frac{1}{\alpha}}, \ \dim M_\alpha = \dim M_{\frac{1}{\alpha}}.$$

Since $\alpha \neq 1$ it follows that $M_\alpha \perp M_{\frac{1}{\alpha}}$. The theorem will be proved if we show that $M_\alpha \oplus M_{\frac{1}{\alpha}} = H$.

From (9.45) it follows that M_α and $M_{\frac{1}{\alpha}}$ are invariant under the U_g. Consequently, $M_\alpha \oplus M_{\frac{1}{\alpha}}$ is invariant under B and the U_g. It therefore follows from Theorem 9.6 that

$$H = M_\alpha \oplus M_{\frac{1}{\alpha}} \tag{9.46}$$

and the theorem is proved.

Corollary 1. *Under the conditions of Theorem* 9.6 *either* $A - A^* = 0$ *or* $A - A^*$ *is a nonsingular operator.*

Corollary 2. *Under the conditions of Theorem* 9.6, *if* $B^* \neq B$ *then* $\dim N_1 = \dim N_{-1}$, *where* N_1 *and* N_{-1} *are the eigenspaces of B corresponding to the eigenvalues 1 and —1.*

Proof. Suppose that $\dim N_1 \neq \dim N_{-1}$. It can be assumed without loss of generality that $\dim N_1 > \dim N_{-1}$. We rewrite (9.46) in the form $M_\alpha \oplus BM_\alpha = H$. From this result we see that $M_\alpha \cap N_1 = 0$. Consequently, $\dim M_\alpha \leqslant \dim N_{-1}$. Hence $\dim H = 2 \dim M_\alpha \leqslant 2 \dim N_{-1} < \dim N_1 + \dim N_{-1} = \dim H$, i.e. we arrive at a contradiction.

Thus $\dim N_1 = \dim N_{-1}$.

Corollary 3. *Under the conditions of Theorem* 9.6, *if* $\dim H = 2n + 1$ $(n = 0, 1, \ldots)$ *then* $B = B^* = B^{-1}$.

Remark. Let $(A, H_{\tilde{U}, U}, \varphi, E_V)$ be an indecomposable operator colligation that is invariant with respect to the group G and whose internal operator is nonsingular. Suppose H is a unitary space and the representations $g \to \tilde{U}_g$, $g \to U_g$ are normal and unitarily equivalent. Then, as follows from Lemma 9.6 and Theorem 9.2, the operator A is described in Theorems 9.6 and 9.7.

§5. PREINVARIANT OPERATOR COLLIGATIONS CONTAINING A GIVEN OPERATOR A AND BIMODULE $H_{\tilde{U},U}$

Let H be a space with metric, let $H_{\tilde{U},U}$ be a bimodule and let A and A^+ be operators acting in $H_{\tilde{U},U}$. Does there exist in this case a space E and an operator φ: $H \to E$ such that $(A, H_{\tilde{U},U}, \varphi, E)$ is a preinvariant operator colligation? We will give an affirmative answer to this question below. In addition, we will completely describe the class of preinvariant operator colligations containing a given operator A and bimodule $H_{\tilde{U},U}$. As follows from Theorem 9.2, we will thereby be giving a complete description of the class of invariant operator colligations containing a given operator A and bimodule $H_{\tilde{U},U}$. We will first prove some auxiliary assertions.

Lemma 9.9. *Any operator A acting in a pseudounitary space H can be included in an operator colligation.*

Proof. Let $H_1 = \operatorname{Ker}(A - A^+)$, let H_2 be a complement of H_1 and let φ be the projection along H_1 onto H_2. We take as E the space H_2 and define a metric in E as follows:

$$(f_1, f_2)_E = \left(\frac{1}{i}(A - A^+) f_1, f_2\right)_H; \; f_1, f_2 \in E.$$

The operator φ can be regarded as an operator acting from H into E. It is easily verified that (A, H, φ, E) is an operator colligation. The lemma is proved.

It follows from the colligation condition (9.1) that if (A, H, φ, E) is an operator colligation then $\dim E \geqslant \dim H - \dim \operatorname{Ker}(A - A^+)$. We will therefore say that an operator colligation is *minimal* if $\dim E = \dim H - \dim \operatorname{Ker}(A - A^+)$. Clearly, a colligation is minimal if and only if $\Delta_\varphi = E$. Thus the proof of Lemma 9.9 provides us with a method of including an arbitrary operator in a minimal colligation.

Lemma 9.10. *Let* $X_i = (A, H, \varphi_i, E_i)$ $(i = 1, 2)$ *be two minimal operator colligations. Then there exists a unitary mapping* V *of* E_1 *onto* E_2 *such that* $V\varphi_1 = \varphi_2$.

Proof. From the colligation condition (9.1) we get $\varphi_1^+\varphi_1 = \varphi_2^+\varphi_2$. Consequently,

$$(\varphi_1 f_1, \ \varphi_1 f_2)_{E_1} = (\varphi_1^+ \varphi_1 f_1, \ f_2)_H = (\varphi_2^+ \varphi_2 f_1, \ f_2)_H = (\varphi_2 f_1, \ \varphi_2 f_2)_{E_2}, \quad f_1, \ f_2' \in H. \tag{9.47}$$

Since $E_i = \Delta_{\varphi_i}$ $(i = 1, \ 2)$ it follows from (9.47) that we can define a unitary operator $V : E_1 \to E_2$ by putting $V\varphi_1 f = \varphi_2 f$, $f \in H$. The lemma is proved.

Theorem 9.8. *Let $\tilde{H}$ be a pseudounitary space, let $H_{\tilde{U}, U}$ be a bimodule over G and let A and A^+ be operators acting in $H_{\tilde{U}, U}$. Then the minimal operator colligation $(A, \ H_{\tilde{U}, U}, \ \varphi, \ E)$ is preinvariant.*

Proof. Since the colligation $(A, \ H_{\tilde{U}, U}, \ \varphi, \ E)$ is minimal, $\Delta_\varphi = E$ and the assertion of the theorem therefore follows from Lemma 9.4.

Let H be a pseudounitary space and let $H_{\tilde{U}, U}$ be a bimodule over G. A subspace $\tilde{H} \subset H$ will be said to be *admissible* if it is a submodule of H_U and its annihilator is a submodule of $H_{\tilde{U}}$. Further, an operator T acting from H into some space E is said to be *admissible* if Ker T is an admissible subspace.

Lemma 9.11. *Suppose $X = (A, \ H_{\tilde{U}, U}, \ \varphi, \ E)$ is an operator colligation and the operators A and A^+ act in the bimodule $H_{\tilde{U}, U}$. Then X is preinvariant if and only if its channel operator φ is admissible.*

Let $X' = (A', \ H'_{\tilde{U}', U'}, \ \varphi', \ E')$ and $X'' = (A'', \ H''_{\tilde{U}'', U''}, \ \varphi'', \ E'')$ be two operator colligations that are preinvariant with respect to the same group G. We will say that *X' stands in the relation M to X''* and write $X'MX''$ if $H' = H''$, $A' = A''$, $\tilde{U}' = \tilde{U}''$, $U' = U''$. The relation M is clearly an equivalence relation. Consequently, the set of operator colligations that are preinvariant with respect to a given group G is partitioned into a set of disjoint M-equivalence classes.

An operation taking a preinvariant operator colligation $(A, \ H_{\tilde{U}, U}, \ \varphi, \ E)$ into a preinvariant colligation $(A, \ H_{\tilde{U}, U}, \ \varphi \dotplus 0, \ E \dotplus E')$, where E' is an arbitrary pseudounitary space, will be called a *trivial lengthening*, while the inverse operation will be called a *trivial shortening*.

Let $X = (A, \ H_{\tilde{U}, U}, \ \varphi, \ E)$ be a preinvariant operator colligation, let E' be a pseudounitary space and let φ' be an admissible operator acting from H into E'. Then the operator colligation $X' = (A, \ H_{\tilde{U}, U}, \ \varphi \dotplus \varphi' \dotplus \varphi', \ E \dotplus E' \dotplus E'^{-})$, where E'^{-} is the space whose metric differs only in sign from the metric in E', is clearly preinvariant. The operation taking X into X' will be called a *neutral lengthening.* If $\varphi' = 0$, the corresponding neutral lengthening will be said to be *trivially neutral.*

An operation taking a preinvariant colligation $(A,\ H_{\tilde{U},U},\ \varphi \dot{+} \varphi' \dot{+} \varphi',\ E \dot{+} E' \dot{+} E'^{-})$, where φ and φ' are admissible operators, into the preinvariant colligation $(A,\ H_{\tilde{U},U},\ \varphi,\ E)$, is called a *neutral shortening.* If $\varphi' = 0$, the corresponding neutral shortening will be said to be *trivially neutral.*

An operation taking a preinvariant colligation $(A,\ H_{\tilde{U},U},\ \varphi,\ E)$ into a preinvariant colligation $(A,\ H_{\tilde{U},U},\ U\varphi,\ E)$, where U is an arbitrary unitary mapping of E onto E', will be called a *unitary operation.*

We note that the M-equivalence classes of operator colligations are invariant under all of the above-mentioned operations. On the other hand, we have

Theorem 9.9. *Let* X_i $(i = 1, 2)$ *be two* M*-equivalent operator colligations. Then one of them can be obtained from the other by means of a finite number of unitary operations, lengthenings and shortenings.*

Proof. From Lemma 9.10 and Theorem 9.8 it follows that it suffices to show that any preinvariant colligation $X = (A,\ H_{\tilde{U},U},\ \varphi,\ E)$ can be obtained from a minimal colligation by means of the above-mentioned operations. Suppose first that Δ_φ is a nondegenerate subspace of E. Then X can be represented in the form

$$X = (A,\ H_{\tilde{U},U},\ \varphi \dot{+} 0,\ \Delta_\varphi \oplus \tilde{E}),\ \tilde{E} = E \ominus \Delta_\varphi . \tag{9.48}$$

Since the colligation $X' = (A,\ H_{\tilde{U},U},\ \varphi,\ \Delta_\varphi)$ is obviously minimal, it follows in the present case, as can be seen from (9.48), that X can be obtained from X' by means of a trivial lengthening.

Suppose now that Δ_φ is a degenerate subspace of E. This means that $\Delta_\varphi \cap \operatorname{Ker} \varphi^+ \neq 0$. Let $\hat{E} = \Delta_\varphi \cap \operatorname{Ker} \varphi^+$. We note that the complete preimage $\varphi^{-1}(\hat{E}) = \tilde{H}$ of E is the kernel of the operator $\varphi^+\varphi$. Since $\tilde{U}_g\varphi^+\varphi = \varphi^+\varphi U_g$, we see that $\tilde{H}$ is invariant under the U_g. Further, $\operatorname{Ker} \varphi \subset \tilde{H}$ and $\operatorname{Ker} \varphi$ is invariant under the U_g. Therefore there exists a subspace $\hat{H}$ invariant under the U_g such that $\tilde{H} = \operatorname{Ker} \varphi \dot{+} \hat{H}$. Let H^c be a complement of $\tilde{H}$ that is invariant under the U_g. Thus $H = \operatorname{Ker} \varphi \dot{+} \hat{H} \dot{+} H^c$. Let $E^c = \varphi H^c$. Then $\Delta_\varphi = \hat{E} \dot{+} E^c$. The subspace E^c is nondegenerate. For suppose $f_0 \in E^c$ and $(f_0, f) = 0$ for any $f \in E^c$. Since $(f_0, \hat{f}) = 0$ for any $\hat{f} \in \hat{E}$, we have $f_0 \in \operatorname{Ker} \varphi^+$. But then $f_0 \in \Delta_\varphi \cap \operatorname{Ker} \varphi^+ = \hat{E}$. Consequently, $f_0 = 0$.

Let L denote the orthogonal complement of E^c, i.e. $E = E^c \oplus L$. Since the vectors of $\hat{E}$ are orthogonal to those of Δ_φ, we have $\hat{E} \subset L$. We represent L in the form $L = L_+ \oplus L_-$, where L_+ and L_- are the subspaces of L such that $(f_+, f_+) > 0$, $(f_-, f_-) < 0$. for any vectors $f_+ \in L_+$, $f_- \in L_-$. Let P_+ be the projection along $L_- \oplus E^c$ onto L_+ and let P_- be the projection along $L_+ \oplus E^c$ onto L_-.

We introduce the operators

$$\varphi_+ h = \begin{cases} P_+\varphi h, & h \in \hat{H}, \\ 0, & h \in \operatorname{Ker}\varphi \dotplus H^c, \end{cases}$$

$$\varphi_- h = \begin{cases} P_-\varphi h, & h \in \hat{H}, \\ 0, & h \in \operatorname{Ker}\varphi \dotplus H^c, \end{cases}$$

$$\varphi_1 = \varphi_+ + \varphi_-, \quad \varphi^c = \varphi - \varphi_1.$$

Clearly,

$$\Delta_{\varphi_1} = \hat{E}, \ \Delta_{\varphi^c} = E^c.$$

If $h_1, h_2 \in \hat{H}$,

$$(\varphi_+h_1, \varphi_+h_2) = (P_+\varphi h_1, P_+\varphi h_2) = (\varphi^+ P_+\varphi h_1, h_2) =$$
$$(\varphi^+ (I - P_-)\varphi h_1, h_2) = (\varphi^+\varphi h_1, h_2) - (P_-\varphi h_1, P_-\varphi h_2).$$

But $h \in H$ implies $\varphi^+\varphi h = 0$. Thus

$$(\varphi_+h_1, \varphi_+h_2) = -(\varphi_-h_1, \varphi_-h_2); \ h_1, h_2 \in \hat{H}. \tag{9.49}$$

Let $\hat{L}_+$ and $\hat{L}_-$ denote the subspaces $P_+\hat{E}$ and $P_-\hat{E}$ respectively and let $L_0 = L \ominus (\hat{L}_+ \oplus \hat{L}_-)$. Clearly, $\Delta_{\varphi_\pm} = \hat{L}_\pm$. From (9.49) we see that there exists a unitary mapping U of $\hat{L}_-$ onto $\hat{L}_+^-$, where $\hat{L}_+^-$ is the space whose metric differs only in sign from the metric in $\hat{L}_+$, such that $U\varphi_- = \varphi_+$. Thus, by carrying out a unitary operation on X, we obtain the colligation

$$X_1 = (A,\ H_{\tilde{U},U},\ \tilde{\varphi} \dotplus \varphi' \dotplus \varphi' \dotplus 0,\ \tilde{E} \dotplus E' \dotplus E'^- \dotplus E_0)$$

where

$$E' = \hat{L}_{+},\ E'^{-} = V\hat{L}_{-},\ E_0 = L_0,\ \tilde{E} = E^c,\ \varphi' = \varphi_{+},\ \tilde{\varphi} = \varphi^c.$$

Clearly, $\tilde{\varphi}^{+}\tilde{\varphi} = \varphi^{+}\varphi = \frac{1}{i}(A - A^{+})$ and $\Delta_{\tilde{\varphi}} = \tilde{E}$. Thus $(A,\ H_{\tilde{U},\,U},\ \tilde{\varphi},\ \tilde{E})$ is a minimal colligation, which we denote by X_2. From Theorem 9.8 it follows that X_2 is preinvariant. The theorem will obviously be proved if we show that XMX_2. To this end it suffices to show that $X_1 M X_2$. If the operator φ' is admissible, X_1 is M-equivalent to X_2 by definition. Otherwise we consider an operator T acting from H into some space Q with metric. Let $\operatorname{Ker} T = \operatorname{Ker} \varphi$. Clearly, T is admissible. Performing a neutral lengthening on X_1 we obtain the preinvariant colligation

$$X_3 = (A,\ H_{\tilde{U},\,U},\ \tilde{\varphi} \dotplus \varphi' \dotplus \varphi' \dotplus 0 \dotplus T \dotplus T,\ \tilde{E} \dotplus E' \dotplus E'^{-} \dotplus E_0 \dotplus \\ \dotplus Q \dotplus Q^{-}).$$

We note that $\operatorname{Ker}(\varphi' \dotplus T) = \operatorname{Ker} \varphi' \cap \operatorname{Ker} T = \operatorname{Ker} \varphi$. Hence the operator $\varphi' \dotplus T$ is admissible. Performing now a unitary operation and a neutral shortening on X_3, we obtain the colligation $X_4 = (A, H_{\tilde{U},\,U},\ \tilde{\varphi} \dotplus 0,\ \tilde{E} \dotplus E_0)$. The theorem follows upon noting that X_4 can be obtained from X_2 by means of a trivial lengthening. We have simultaneously obtained the following result.

Let $X = (A,\ H_{\tilde{U},\,U},\ \varphi,\ E)$ be a preinvariant colligation and suppose Δ_φ is a degenerate subspace of E. Then X can be reduced by a unitary operation, a neutral lengthening and a neutral shortening to the form

$$(A,\ H_{\tilde{U},\,U},\ \tilde{\varphi} \dotplus 0,\ \tilde{E} \dotplus E_0),$$

where $(A,\ H_{\tilde{U},\,U},\ \tilde{\varphi},\ \tilde{E})$ is a minimal colligation and

$$\dim(\tilde{E} \dotplus E_0) < \dim E.$$

Let H be a pseudounitary space, let $H_{\tilde{U},\,U}$ be a bimodule and let A and A^{+} be operators acting in $H_{\tilde{U},\,U}$. From the arguments carried out above it follows that all of the preinvariant operator colligations containing a given operator A and bimodule $H_{\tilde{U}, U}$ can be described by including A in a minimal colligation and making use of Theorems 9.8 and 9.9.

We note that in proving Theorem 9.9 we have proved the following assertion.

Theorem 9.10. *Any preinvariant colligation* $(A, \ H_{\tilde{U}, U}, \varphi, E)$ *can be reduced by means of a unitary operation to the form*

$$(A, \ H_{\tilde{U}, U}, \ \tilde{\varphi} \dotplus \varphi' \dotplus \varphi' \dotplus 0, \ \tilde{E} \dotplus E' \dotplus E'^{-} \dotplus E_0), \qquad (9.50)$$

where $(A, \ H_{\tilde{U}, U}, \ \tilde{\varphi}, \ \tilde{E})$ *is a minimal colligation,* $\operatorname{Ker} \varphi' = \operatorname{Ker} \varphi + H^c$ *is invariant under the* U_g, E' *is a unitary space and* $\Delta_{\varphi'} = E'$.

Corollary. *If the internal representations of a preinvariant colligation* $X = (A, \ H_{\tilde{U}, U}, \ \varphi, \ E)$ *satisfy the condition*

$$\tilde{U}_g^{+} = U_g^{-1}, \qquad (9.51)$$

the operator φ' *in the representation* (9.50) *is admissible.*

Proof. In fact, in the present case any subspace that is invariant under the U_g is admissible.

Theorem 9.11. *Suppose the internal representations of a colligation* $X = (A, \ H_{\tilde{U}, U}, \ \varphi, \ E)$ *that is preinvariant with respect to a group* G *satisfy condition* (9.51). *Then a unitary representation* $g \to V_g$, $g \in G$, *such that the colligation* $(A, \ H_{\tilde{U}, U}, \ \varphi, \ E_V)$ *will be invariant with respect to* G *can be defined on* E.

Proof. We represent the colligation X in the form (9.50). From Lemma 9.3 it follows that a representation $g \to \tilde{V}_g$, such that $\tilde{V}_g\tilde{\varphi} = \tilde{\varphi}U_g$ can be defined on $\tilde{E}$.

Let $\varphi_1 = \varphi' \dotplus \varphi'$, $E_1 = \Delta_{\varphi_1}$, P_+ be the projection along $\tilde{E} \dotplus E'^{-} \dotplus E_0$ onto E', P_- be the projection along $\tilde{E} \dotplus E' \dotplus E_0$ onto E'^{-}, $J = P_+ - P_-$ and $E_2 = JE_1$. From the Corollary of Theorem 9.10 we see that the operator φ_1 is admissible. It therefore follows from Lemma 9.3 that a representation $g \to V_g^{(1)}$ such that $V_g^{(1)}\varphi_1 = \varphi_1 U_g$ can be defined on E_1.

Suppose $f_1, f_2 \in E_1$ and h_1, h_2 are vectors in H such that $\varphi_1 h_i = f_i$. Since $\varphi_1^+\varphi_1 = 0$, we have

$$(f_1, f_2)_E = (\varphi_1 h_1, \varphi_1 h_2)_E = (\varphi_1^+\varphi_1 h_1, h_2)_H = 0.$$

This implies that $E_1 \subset E_1^0$, where E_1^0 denotes the annihilator of E_i. Hence $\tilde{E}_1 \dotplus E_1 \dotplus E_0 \subseteq E_1^0$. Noting that

$$\dim(\tilde{E} \dotplus E_1 \dotplus E_0) = \dim E - \dim E_1 = \dim E_1^0,$$

we get $E_1^0 = \tilde{E} \dotplus E_1 \dotplus E_0$. It is not difficult to see that E_2 is a complement of E_1^0. Therefore φ_1^+ is nonsingular on E_2 and $\Delta_{\varphi_1}^+ = \varphi_1^+ E_2$. Since φ_1 is an admissible operator, $\varphi_1^+ E_2$ is invariant under the $\tilde{U}_g$. Thus a representation $g \to V_g^{(2)}$ can be defined on E_2 by putting

$$\tilde{U}_g \varphi_1^+ f = \varphi_1^+ V_g^{(2)} f, \quad f \in E_2.$$

We define a representation $g \to V_g$ on E as follows:

$$V_g = \tilde{V}_g \tilde{P} + V_g^{(1)} P_1 + V_g^{(2)} P_2 + P_0,$$

where $\tilde{P}$ is the projection along $E_1 \dotplus E_2 \dotplus E_0$ onto $\tilde{E}$ and P_i is the projection along $\tilde{E} \dotplus E_j \dotplus E_k$ onto E_i $(i, j, k = 0, 1, 2;\ i \neq j,\ i \neq k,\ j \neq k)$.

Let us show that the representation $g \to V_g$ is a unitary representation. To this end we first note that the representation $g \to \tilde{V}_g$ is unitary on $\tilde{E}$. In fact, any vector $f \in \tilde{E}$ can be represented in the form $f = \tilde{\varphi} h$, $h \in H$. Therefore, for any $f_1, f_2 \in \tilde{E}$ we get

$$(\tilde{V}_g f_1, \tilde{V}_g f_2)_E = (\tilde{V}_g \tilde{\varphi} h_1, \tilde{V}_g \tilde{\varphi} h_2)_E = (\tilde{\varphi} U_g h_1, \tilde{\varphi} U_g h_2)_E =$$
$$(\tilde{\varphi}^+ \tilde{\varphi} U_g h_1, U_g h_2)_H = (\tilde{U}_g \tilde{\varphi}^+ \tilde{\varphi} h_1, U_g h_2)_H = (\tilde{\varphi}^+ \tilde{\varphi} h_1, h_2)_H = (f_1, f_2)_E,$$

where $f_i = \tilde{\varphi} h_i$, $h_i \in H$.

We now prove that the representation $g \to \hat{V}_g = V_g^{(1)} P_1 + V_g^{(2)} P_2$ is unitary on $E_1 + E_2$. Let $e,\ e' \in E_1 + E_2$. Then

$$(\hat{V}_g e,\ \hat{V}_g e') = (\hat{V}_g (P_1 e + P_2 e),\ \hat{V}_g (P_1 e' + P_2 e')) =$$
$$(V_g^{(1)} P_1 e,\ V_g^{(1)} P_1 e') + (V_g^{(1)} P_1 e,\ V_g^{(2)} P_2 e') +$$
$$(V_g^{(2)} P_2 e,\ V_g^{(1)} P_1 e') + (V_g^{(2)} P_2 e',\ V_g^{(2)} P_2 e').$$

Since the subspaces E_1 and E_2 are contained in their annihilators,

$$(V_g^{(1)} P_1 e,\ V_g^{(1)} P_1 e') = 0, \quad (V_g^{(2)} P_2 e,\ V_g^{(2)} P_2 e') = 0.$$

Consequently,

$$(\hat{V}_g e,\ \hat{V}_g e') = (V_g^{(1)} P_1 e,\ V_g^{(2)} P_2 e') + (V_g^{(2)} P_2 e,\ V_g^{(1)} P_1 e').$$

The vector $P_1 e$ can obviously be represented in the form $P_1 e = \varphi_1 h,\ h \in H$. Hence

$$(V_g^{(1)} P_1 e,\ V_g^{(2)} P_2 e')_E = (V_g^{(1)} \varphi_1 h,\ V_g^{(2)} P_2 e')_E =$$
$$(\varphi_1 U_g h,\ V_g^{(2)} P_2 e')_E = (U_g h,\ \varphi^+ V_g^{(2)} P_2 e')_H =$$
$$(U_g h,\ \tilde{U}_g \varphi^+ P_2 e')_H = (h,\ \varphi^+ P_2 e')_H = (P_1 e,\ P_2 e')_E.$$

It can be shown in the same way that

$$(V_g^{(2)} P_2 e,\ V_g^{(1)} P_1 e') = (P_2 e,\ P_1 e').$$

Consequently,

$$(\hat{V}_g e,\ \hat{V}_g e') = (P_1 e,\ P_2 e') + (P_2 e,\ P_1 e') = (e,\ e').$$

The unitariness of the representation $g \to V_g$ now follows upon noting that the subspaces $\tilde{E}$, $E_1 + E_2$ and E_0 are orthogonal to each other.

Finally, we show that the colligation $(A, H_{\tilde{U},U}, \varphi, E_V)$ is invariant with respect to the group G. From the definition of the representation $g \to V_g$ we have $V_g\varphi = \varphi U_g$. Hence

$$\varphi^+ V_g^+ = U_g^+ \varphi^+. \tag{9.52}$$

Since $V_g^+ = V_g^{-1}$, $U_g^+ = U_g^{-1}$, (9.52) implies $\tilde{U}_g\varphi^+ = \varphi^+ V_g$, and the theorem is proved.

Remark. Under the conditions and assumptions of Theorem 9.11 a unitary representation can be taken for the external representation $g \to V_g$, in which case the following two conditions are satisfied.

1) The space E decomposes into a direct sum $E = E_0^c \oplus E_0$, in which the subspace E_0^c $(= \tilde{E} \dotplus E_1 \dotplus E_2)$ reduces the operators V_g. In addition, on E_0^c the representation $g \to V_g$ splits into a direct sum of irreducible representations, each of which enters into either the representation $g \to U_g$ or the representation $g \to \tilde{U}_g$, while on E_0 one has $V_g = I$.

2) The subspace E_0^c reduces the c.o.f. $S_X(\lambda)$ of X, and in this connection $S_X(\lambda)|_{E_0} = I$.

Remark. When $\tilde{U} = U$, condition (9.51) means that the internal representation is a unitary representation.

CHAPTER X

WEYL FAMILIES OF OPERATOR COLLIGATIONS AND THEIR CORRESPONDING OPEN FIELDS

§1. BASIC CONCEPTS

All of the spaces considered in this chapter are assumed to be finite dimensional linear spaces over the field of complex numbers.

1. Suppose we are given a group G of coordinate transformations in a fixed space P and a linear representation $g \to U_g$ of G on a space H. We recall that by a *variable* corresponding to the representation $g \to U_g$ is meant a vector $h \in H$ referred to a given coordinate system in P and going over into $U_g h$ under a coordinate transformation g in P.

Suppose G is the Lorentz group, $\| g_{jk} \|_{j,k=0}^{3}$ is the corresponding matrix of an element $g \in G$ and $g \to U_g$ is a representation of G on a space H. We will consider the differential equation

$$\frac{1}{i} \sum_{k=0}^{n} L_k \frac{\partial h}{\partial x_k} + \chi h = 0, \qquad (10.1)$$

where h is an unknown function of the x_k ($k = 0, 1, 2, 3$) whose values are variables corresponding to the representation $g \to U_g$, the L_k are given linear operators in H and χ is a real number.

Equation (10.1) is called an *invariant equation* with respect to the Lorentz group if it does not change under a transformation $g \in G$ of the point (x_0, x_1, x_2, x_3) and simultaneous transformation U_g of the variable h.

We recall that a necessary and sufficient condition for the invariance of equation (10.1) with respect to the Lorentz group is the fulfillment of the conditions

$$\sum_{k=0}^{3} g_{jk} \tilde{U}_g L_k U_g^{-1} = L_j \quad (j = 0, 1, 2, 3) \tag{10.2}$$

where $g \to \tilde{U}_g$ is also a representation of the Lorentz group on H; in this connection, $\tilde{U}_g = U_g$ for all $g \in G$ if $\chi \neq 0$.[1]

Suppose equation (10.1) is invariant with respect to the Lorentz group and H is a pseudounitary space. Then equation (10.1) can be obtained from an invariant Lagrangian if and only if

$$L_k^+ = L_k \quad (k = 0, 1, 2, 3), \tag{10.3}$$

$$\tilde{U}_g^+ = U_g^{-1}. \tag{10.4}$$

Definition. Let H be a pseudounitary space, let $H_{\tilde{U},U}$ be a bimodule over G and let $\{L_k\}_{k=0}^{3}$ be operators acting in $H_{\tilde{U},U}$. We will say that

$$(\{L_k\}_{k=0}^{3}, H_{\tilde{U},U}) \tag{10.5}$$

is a *connective* if conditions (10.2)–(10.4) are satisfied.

Thus, to each equation (10.1) that is (i) invariant with respect to the Lorentz group and (ii) obtainable from an invariant Lagrangian, we can put in correspondence a connective (10.5), and conversely; in this connection, if $\tilde{U} \neq U$ there corresponds to the connection (10.5) an equation (10.1) with $\chi = 0$.

[1] A detailed account of the theory of equations that are invariant with respect to the Lorentz group is given in [14, 32].

We recall that an equation (10.1) that is invariant with respect to the Lorentz group is said to be *indecomposable* if there does not exist a nondegenerate subspace in H that reduces the operators $\{L_k\}_{k=0}^{3}$ and decomposes the bimodule $H_{\tilde{U}, U}$. In connection with this we introduce the following

Definition. A connective (10.5) will be said to be *indecomposable* if the equation (10.1) corresponding to this connective is indecomposable.

In relativistic quantum mechanics it is assumed that the field of a free particle should be described by an indecomposable equation that is invariant with respect to the Lorentz group and obtainable from a Lagrangian. Thus, to each free particle we can put in correspondence an indecomposable connective. Following the formalism adopted in physics, we can assume that a free particle corresponds to each indecomposable equation (10.1) that is invariant and obtainable from an invariant Lagrangian.

2. Let $(\{L_k\}_{k=0}^{3}, H_{\tilde{U}, U})$ be a connective. If this connective decomposes, the equation (10.1) corresponding to it also decomposes into some indecomposable equations. This corresponds to the fact that equation (10.1) describes a field of free particles in the case under consideration.

Let $X = (A, H, \varphi, E)$ be an operator colligation and let $g \to V_g$ be a representation of a group G on E. The basic idea in effecting a passage to open fields is to consider the equations

$$\begin{gathered} \frac{1}{i}\sum_{k=0}^{3} L_k \frac{\partial h}{\partial x_k} + Ah = \varphi^+ u; \\ v = u - i\varphi h, \end{gathered} \tag{10.6}$$

where u and v are functions of x_0, x_1, x_2, x_3 whose values are variables corresponding to the representation $g \to V_g^*$.[2]

The operator A can be regarded as an operator describing the linear interactions between the particles.

We will deduce conditions for the invariance of equations (10.6) with respect to the Lorentz group.

[2]Equations (10.6) have the structure of open system equations. We note in connection with this that a detailed account of the theory of open systems is given in [27[1]].

We subject $x = (x_0, x_1, x_2, x_3)$ to a Lorentz transformation $g : x' = gx$, i.e.

$$x'_j = \sum_{k=0}^{3} g_{jk} x_k \, (j = 0, 1, 2, 3).$$

In this connection h is transformed by the operator U_g while u and v are transformed by the operator V_g:

$$h'(x') = U_g h(x), \quad u'(x') = V_g u(x), \quad v'(x') = V_g v(x).$$

Then $h = U_g^{-1} h', \quad u = V_g^{-1} u', \quad v = V_g^{-1} v'$ and

$$\frac{\partial}{\partial x_k} = \sum_{j=0}^{3} g_{jk} \frac{\partial}{\partial x'_j} \, (k = 0, 1, 2, 3).$$

Substituting these expressions into equations (10.6), we get

$$\frac{1}{i} \sum_{j=0}^{3} \sum_{k=0}^{3} g_{jk} L_k U_g^{-1} \frac{\partial h'}{\partial x'_j} + A U_g^{-1} h' = \varphi^+ V_g^{-1} u',$$
$$V_g^{-1} v' = V_g^{-1} u' - i \varphi U_g^{-1} h'.$$

Hence

$$\frac{1}{i} \left(\sum_{j=0}^{3} \sum_{k=0}^{3} g_{jk} \tilde{U}_g L_k U_g^{-1} \right) \frac{\partial h'}{\partial x'_j} + \tilde{U}_g A U_g^{-1} h' = \tilde{U}_g \varphi^+ V_g^{-1} u', \qquad (10.7)$$
$$v' = u' - i V_g \varphi U_g^{-1} h'.$$

Since equations (10.6) are assumed to be invariant, they must coincide with equations (10.7), i.e. the following conditions must be satisfied:

$$\sum_{k=0}^{3} g_{jk} \tilde{U}_g L_k U_g^{-1} = L_j \quad (j = 0, 1, 2, 3); \qquad (10.8)$$

$$\tilde{U}_g A U_g^{-1} = A; \qquad (10.9)$$

$$\tilde{U}_g \varphi^+ V_g^{-1} = \varphi^+; \tag{10.10}$$

$$V_g \varphi U_g^{-1} = \varphi. \tag{10.11}$$

But conditions (10.8) are satisfied, since $(\{L_k\}_{k=0}^3,\ H_{\tilde{U}}, U)$ is a connective. Therefore equations (10.6) are invariant if and only if the invariance conditions (10.9)–(10.11) for the colligation X are satisfied.

Definition. Equations of the form

$$\begin{gathered} \frac{1}{i}\sum_{k=0}^{3} L_k \frac{\partial h(x)}{\partial x_k} + Ah(x) = \varphi^+ u(x), \\ v(x) = u(x) - i\varphi h(x), \end{gathered} \tag{10.12}$$

where $x = (x_0,\ x_1,\ x_2,\ x_3)$, $(\{L_k\}_{k=0}^3,\ H_{\tilde{U}}, U)$ is a connective and $(A,\ H_{\tilde{U}}, U,\ \varphi, E_V)$ is an operator colligation that is invariant with respect to the Lorentz group, will be called *open field equations.*

As in the theory of open systems, we will call $u(x)$ the *input*, $h(x)$ the *internal state* and $v(x)$ the *output* of the *open field* (10.12).

3. We will solve the open field equations (10.12) for the case when the input has the form

$$u(x_0,\ x_1,\ x_2,\ x_3) = u(p_0,\ p_1,\ p_2,\ p_3)\, e^{i(-p_0x_0+p_1x_1+p_2x_2+p_3x_3)}$$

(such an input will be called a *plane wave*). The variable $u(p_0, p_1, p_2, p_3) = u(p)$ does not depend on $x_0,\ x_1,\ x_2,\ x_3$ and the numbers $p_0,\ p_1,\ p_2,\ p_3$ are assumed to be real. It is not difficult to show that the vector $u(p_0,\ p_1,\ p_2,\ p_3)$ is transformed according to the rule $u(p') = V_g u(p)$ where $p' = gp$, i.e.

$$p_i' = \sum_{k=0}^{3} g_{ik} p_k \quad (i = 0,\ 1,\ 2,\ 3).$$

We naturally seek a solution of the equation

$$\frac{1}{i}\sum_{k=0}^{3} L_k \frac{\partial h(x)}{\partial x} + Ah(x) = \varphi^+ u(p)\, e^{i(-p_0x_0+p_1x_1+p_2x_2+p_3x_3)} \tag{10.13}$$

in the form

$$h(x_0, x_1, x_2, x_3) = h(p_0, p_1, p_2, p_3)\, e^{i(-p_0x_0+p_1x_1+p_2x_2+p_3x_3)}, \quad (10.14)$$

where the variable $h(p_0, p_1, p_2, p_3)$ no longer depends on x_0, x_1, x_2, x_3. Substituting the function (10.14) into equation (10.13), we get

$$(-L_0p_0 + \sum_{\alpha=1}^{3} L_\alpha p_\alpha)\, h(p) + Ah(p) = \varphi^+ u(p).$$

If the operator $L_0p_0 + \sum_{\alpha=1}^{3} L_\alpha p_\alpha + A$ has an inverse, then $h(p) = R(p)\, u(p)$, where $R(p) = (-L_0p_0 + \sum_{\alpha=1}^{3} L_\alpha p_\alpha + A)^{-1}\varphi^+$.

It is not difficult to see that the output $v(x)$ will have the form

$$v(x_0, x_1, x_2, x_3) = v(p_0, p_1, p_2, p_3)\, e^{i(-p_0x_0+p_1x_1+p_2x_2+p_3x_3)},$$

where the variable $v(p_0, p_1, p_2, p_3) = v(p)$ does not depend on x_0, x_1, x_2, x_3 and $v(p) = u(p) - i\varphi h(p)$.

It can be shown that the variables $h(p)$ and $v(p)$ are transformed according to the rule

$$h(p') = U_g h(p), \quad v(p') = V_g v(p), \quad p' = gp.$$

We note that $v(p)$ is connected with $u(p)$ in the following manner:

$$v(p) = [I - i\varphi(-L_0p_0 + \sum_{\alpha=1}^{3} L_\alpha p_\alpha + A)^{-1}\varphi^+]\, u(p).$$

Definition. The operator function

$$S(p_0, p_1, p_2, p_3) = I - i\varphi(-L_0p_0 + \sum_{\alpha=1}^{3} L_\alpha p_\alpha + A)^{-1}\varphi^+ \quad (10.15)$$

will be called the *characteristic operator function* (c.o.f.) of the *open field* (10.12).[3]

[3]Cf. Chapter IV, §1.

Thus an open field in the case when the input is a plane wave is completely determined by the mappings

$$R(p): E \to H, \; S(p): E \to E.$$

Solving the open field equations (10.12) for the case when the input is a plane wave, we obtain the family of operator colligations

$$\left(-L_0 p_0 + \sum_{\alpha=1}^{3} L_\alpha p_\alpha + A, \; H_{\tilde{U},U}, \; \varphi, \; E_V\right),$$

where $(A, H_{\tilde{U},U}, \varphi, E_V)$ is an invariant colligation with respect to the Lorentz group and $(\{L_k\}_{k=0}^{3}, H_{\tilde{U},U})$ is a connective.

Definition. The family of operator colligations

$$X(p) = \left(-L_0 p_0 + \sum_{\alpha=1}^{3} L_\alpha p_\alpha + A, \; H_{\tilde{U},U}, \; \varphi, \; E_V\right), \qquad (10.16)$$

where $p = (p_0, p_1, p_2, p_3) \in R^4$, will be called an *invariant family of operator colligations* over the Lorentz group if $(A, H_{\tilde{U},U}, \varphi, E_V)$ is invariant with respect to the Lorentz group and $(\{L_k\}_{k=0}^{3}, H_{\tilde{U},U})$ is a connective.

Thus to each open field there corresponds a well-defined invariant family of operator colligations. And conversely, to each invariant family of operator colligations (10.16) there obviously corresponds an open field (10.12).

We note that if we have an invariant family of operator colligations (10.16), we can consider the operator function (10.15) independently of the notion of an open field. In this case the function (10.15) will be called the *characteristic operator function* of an invariant family of operator colligations.

Remark. All of the definitions involving invariant colligations carry over to invariant families of colligations. For example, the operator A of an invariant family of colligations (10.16) will be called the *internal operator* while the operator φ will be called the *channel operator.*

§2. DECOMPOSITION OF AN INVARIANT FAMILY OF COLLIGATIONS

1. Let

$$X_i(p) = \left(-L_0^{(i)} p_0 + \sum_{\alpha=1}^{3} L_\alpha^{(i)} p_\alpha + A_i,\ H^{(i)}_{\tilde{U}^{(i)},\,U^{(i)}},\ \varphi_i,\ E_V\right)\ (i = 1, 2) -$$

be two invariant families of operator colligations and let P_1 and P_2 denote the orthogonal projections onto $H^{(1)}$ and $H^{(2)}$ in the space $H = H^{(1)} \oplus H^{(2)}$. Thus P_1 is the projection along $H^{(2)}$ onto $H^{(1)}$ while P_2 is the projection along $H^{(1)}$ onto $H^{(2)}$. Consider the family of colligations

$$X(p) = \left(-L_0 p_0 + \sum_{\alpha=1}^{3} L_\alpha p_\alpha + A,\ H_{\tilde{U},\,U},\ \varphi,\ E_V\right)$$

where

$$L_j = L_j^{(1)} P_1 + L_j^{(2)} P_2 \quad (j = 0,\ 1,\ 2,\ 3),$$
$$A = A_1 P_1 + A_2 P_2 + i\varphi_2^+ \varphi_1 P_1, \quad \varphi = \varphi_1 P_1 + \varphi_2 P_2,$$
$$\tilde{U}_g = \tilde{U}_g^{(1)} P_1 + \tilde{U}_g^{(2)} P_2, \quad U_g = U_g^{(1)} P_1 + U_g^{(2)} P_2.$$

The family of colligations $X(p)$ is clearly invariant. We will call it the *product* of the families of colligations $X_1(p)$ and $X_2(p)$ and write $X(p) = X_1(p) \vee X_2(p)$.

Direct verification shows that

$$(X_1(p) \vee X_2(p)) \vee X_3(p) = X_1(p) \vee (X_2(p) \vee X_3(p)). \quad (10.17)$$

Let

$$X(p) = \left(-L_0 p_0 + \sum_{\alpha=1}^{3} L_\alpha p_\alpha + A,\ H_{\tilde{U},\,U},\ \varphi,\ E_V\right)$$

be an invariant family of colligations and suppose a nondegenerate subspace H' effects a decomposition of the connective $(\{L_k\}_{k=0}^{3},\ H_{\tilde{U},\,U})$, i.e. reduces the operators $\{L_k\}_{k=0}^{3}$ and decomposes the bimodule $H_{\tilde{U},\,U}$. Let P' denote the projection along $H \ominus H'$ onto H' and consider the family of colligations

$$X(p) = \left(-L_0' p_0 + \sum_{\alpha=1}^{3} L_\alpha' p_\alpha + A',\ H'_{\tilde{U}',\,U'},\ \varphi',\ E_V\right)$$

where

$$L'_j = L_{j\,|\,H'}\ (j = 0,\ 1,\ 2,\ 3),\quad A' = P'A|_{H'}$$
$$\varphi' = \varphi_{|H'},\quad \tilde{U}'_g = \tilde{U}_{g\,|\,H'},\quad U'_g = U_g\,|_{H'}.$$

Clearly, the family of colligations $X'(p)$ is invariant. We will call it the *projection* of $X(p)$ onto $H'$and write $X'(p) = \Pr_{H'} X(p)$.

If

$$X_i(p) = (-L_0^{(i)} p_0 + \sum_{\alpha=1}^{3} L_\alpha^{(i)} p_\alpha + A_i,\ H^{(i)}_{\tilde{U}^{(i)},\,U^{(i)}},\ \varphi_i,\ E_V)\quad (i = 1,\ 2)$$

are two invariant families of colligations and

$$X(p) = X_1(p) \vee X_2(p) = (-L_0 p_0 + \sum_{\alpha=1}^{3} L_\alpha p_\alpha + A,\quad H_{\tilde{U},\,U},\ \varphi,\ E_V),$$

then, as follows from formulas (10.17), the families of colligations $X_1(p)$ and $X_2(p)$ are the projections of the invariant family of colligations $X(p)$ onto $H^{(1)}$ and $H^{(2)}$ respectively, with $H^{(2)}$ being invariant under A. Conversely, each invariant family of colligations

$$X(p) = (-L_0 p_0 + \sum_{\alpha=1}^{3} L_\alpha p_\alpha + A,\ H_{\tilde{U},\,U},\ \varphi,\ E_V)$$

is the product of its projections

$$X_1(p) = (-L_0^{(1)} p_0 + \sum_{\alpha=1}^{3} L_\alpha^{(1)} p_\alpha + A_1,\ H^{(1)}_{\tilde{U}^{(1)} U^{(1)}},\ \varphi_1,\ E_V)$$

and

$$X_2(p) = (-L_0^{(2)} p_0 + \sum_{\alpha=1}^{3} L_\alpha p_\alpha + A_2,\ H^{(2)}_{\tilde{U}^{(2)},\,U^{(2)}},\ \varphi_2,\ E_V)$$

onto the subspace $H^{(1)}$, which effects a decomposition of the connective $(\{L_k\}_{k=0}^{3},\ H_{\tilde{U},U})$ and is invariant under A, and the orthogonal complement $H^{(2)} = H \ominus H^{(1)}$, respectively.

Definition. An invariant family of colligations will be said to be *decomposable* if it can be represented in the form of a product of invariant families of operator colligations.

2. Let $F_i\ (i=1,\ 2,\ 3)$ be an open field to which there corresponds an invariant family of operator colligations

$$X_i(p)=\left(-L_0^{(i)}p_0+\sum_{\alpha=1}^{3}L_\alpha^{(i)}p_\alpha+A_i,\ H_{\tilde{U}^{(i)},\,U^{(i)}}^{(i)},\ \varphi_i,\ E_V\right)$$

and which in the case of a plane wave input is determined by a pair of mappings

$$R_i(p):E\to H^{(i)},\quad S_i(p):E\to E.$$

Definition. We will say that the field F_1 is a *coupling* of the fields F_2 and F_3 and write $F_1=F_2\vee F_3$, if

$$\begin{gathered}H^{(1)}=H^{(2)}\oplus H^{(3)},\quad L_k^{(1)}=L_k^{(2)}P_2+L_k^{(3)}P_3\ (k=0,\ 1,\ 2,\ 3)\\ \tilde{U}_g^{(1)}=\tilde{U}_g^{(2)}P_2+\tilde{U}_g^{(3)}P_3,\quad U_g^{(1)}=U_g^{(2)}P_2+U_g^{(3)}P_3;\\ R_1(p)=R_2(p)+R_3(p)\,S_2(p);\\ S_1(p)=S_3(p)\,S_2(p),\end{gathered}$$

Here P_i is the projection in $H^{(1)}$ along $H^{(i)}$ onto $H^{(j)}\ (i,\ j=2,\ 3;\ i\neq j)$.

As in Chapter IX, we can prove the following assertion.

Theorem 10.1. *Let F be an open field to which there corresponds an invariant family of colligations $X(p)$. If the family of colligations $X(p)$ admits a decomposition into invariant families of colligations $X_1(p)$ and $X_2(p)$ such that $X(p)=X_1(p)\vee X_2(p)$, then $F=F_1\vee F_2$, where $F_i\ (i=1,\ 2)$ is the open field corresponding to the family of colligations $X_i(p)$.*

§3. WEYL FAMILIES OF OPERATOR COLLIGATIONS

1. Let $\{\sigma_k\}_{k=0}^{3}$ be the so-called *Pauli matrices*, i.e. let

$$\sigma_0=\begin{pmatrix}1&0\\0&1\end{pmatrix},\quad \sigma_1=\begin{pmatrix}0&1\\1&0\end{pmatrix},\quad \sigma_2=\begin{pmatrix}0&-i\\i&0\end{pmatrix},\quad \sigma_3=\begin{pmatrix}1&0\\0&-1\end{pmatrix}.$$

The Pauli matrices satisfy the relations

$$\sum_{k=0}^{3} g_{jk}\tilde{c}_g\sigma_k c_g^{-1} = \sigma_j \quad (j=0,\ 1,\ 2,\ 3),$$

where $\tilde{c}: g \rightarrow \tilde{c}_g$ and $c: g \rightarrow c_g$ are the representations of the proper Lorentz group respectively defined by the pairs $\left(\frac{1}{2},\ -\frac{3}{2}\right)$ and $\left(\frac{1}{2},\ \frac{3}{2}\right)$, i.e. $\tilde{c}_g^{*} = \tilde{c}_g^{-1}$[4]. Thus $(\{\sigma_k\}_{k=0}^{3},\ M_{\tilde{c},\,c})$, where M is a two-dimensional unitary space, is a connective.

It is known[5] that the matrices $\{\tilde{\sigma}_k\}_{k=0}^{3}$, where $\tilde{\sigma}_0 = \sigma_0$, $\sigma_\alpha = -\sigma_\alpha$ $(\alpha = 1,\ 2,\ 3)$, satisfy the conditions

$$\sum_{k=0}^{3} g_{jk}c_g\tilde{\sigma}_k c_g^{-1} = \sigma_j \quad (j=1,\ 2,\ 3),$$

i.e. $(\{\tilde{\sigma}_k\}_{k=0}^{3},\ M_{c,\,\tilde{c}})$, where M is a two-dimensional unitary space, is a connective.

Consider now in a unitary space H the operators $\{N_k\}_{k=0}^{3}$ whose matrices relative to some orthonormal basis are

$$\begin{pmatrix} \underbrace{\begin{matrix} \sigma_k & & \\ & \ddots & \\ & & \sigma_k \end{matrix}}_{n} & \\ & \underbrace{\begin{matrix} \tilde{\sigma}_k & & \\ & \ddots & \\ & & \tilde{\sigma}_k \end{matrix}}_{m} \end{pmatrix} \quad (k=0,\ 1,\ 2,\ 3). \qquad (10.18)$$

Let us agree to refer to the orthonormal basis of H relative to which the matrices of the operator $\{N_k\}_{k=0}^{3}$ have the form (10.18) as the *canonical basis.*

We define a pair of representations $\tilde{U}: g \rightarrow \tilde{U}_g$, $U: g \rightarrow U_g$ of the proper Lorentz group on H by putting

---

[4]See, for example, [14].

[5]See, for example, [14].

$$\tilde{U}_g = \begin{pmatrix} \underbrace{Cg^{*-1} \cdots Cg^{*-1}}_{n} & \\ & \underbrace{Cg \cdots Cg}_{m} \end{pmatrix}, \quad U_g = \begin{pmatrix} \underbrace{Cg \cdots Cg}_{n} & \\ & \underbrace{Cg^{*-1} \cdots Cg^{*-1}}_{m} \end{pmatrix} \tag{10.19}$$

relative to the canonical basis. Clearly, $(\{N_k\}_{k=0}^3, \ H_{\tilde{U},U})$ is a connective.

Definition. An invariant family of colligations

$$X(p) = (-N_0 p_0 + \sum_{\alpha=1}^{3} N_\alpha p_\alpha + A, \ H_{\tilde{U},U}, \ \varphi, \ E_V)$$

where H is a unitary space and the operators $\{N_k\}_{k=0}^3, \ \tilde{U}_g, \ U_g$ are defined by formulas (10.18), (10.19) relative to an orthonormal basis, will be called a *Weyl family*.

2. In this subsection we will describe indecomposable Weyl families of colligations. We first note without proof the following simple assertion.

Lemma 10.1. The internal representations $g \to \tilde{U}_g, \ g \to U_g$ of a Weyl family of colligations are normal and their invariant subspaces coincide and reduce the operators $\{N_k\}_{k=0}^3$.

Lemma 10.1 directly implies

Theorem 10.2. *A Weyl family of colligations* $X(p) = (-N_0 p_0 + \sum_{\alpha=1}^{3} N_\alpha p_\alpha + A, \ H_{\tilde{U},U}, \ \varphi, \ E_V)$ *is decomposable if and only if the invariant colligation* $(A, \ H_{\tilde{U},U}, \ \varphi, \ E_V)$ *decomposes into invariant colligations.*

Let $X(p) = (-N_0 p_0 + \sum_{\alpha=1}^{3} N_\alpha p_\alpha + A, \ H_{\tilde{U},U}, \ \varphi, \ E_V)$ be an indecomposable Weyl family of colligations. Suppose first that $\operatorname{Ker} A \neq 0$. It is not difficult to see that in this case $A = 0$ and the representations $g \to \tilde{U}_g, \ g \to U_g$ do not have a common nontrivial invariant subspace. This is clearly possible when, in expression (10.19), either $n = 1, \ m = 0$ or $n = 0, \ m = 1$. In the first case

$$X(p) = (-\sigma_0 p_0 + \sum_{\alpha=1}^{3} \sigma_\alpha p_\alpha, \ H_{\tilde{U},U}, \ \varphi, \ E_V);$$

here $\tilde{U}_g = c_g^{*-1}$, $U_g = c_g$. In the second case

$$X(p) = \left(-\tilde{\sigma}_0 p_0 + \sum_{\alpha=1}^{3} \tilde{\sigma}_\alpha p_\alpha,\quad H_{\tilde{U},U},\ \varphi,\ E_V\right)$$

and $\tilde{U}_g = c_g$, $U_g = c_g^{*-1}$. Also, as is easily seen, in both cases either $\Delta_{\varphi+} = 0$ or $\Delta_{\varphi+} = H$. We have consequently proved

Theorem 10.3. *Let* $X(p) = \left(-N_0 p_0 + \sum_{\alpha=1}^{3} N_\alpha p_\alpha + A,\ H_{\tilde{U},U},\ \varphi,\ E_V\right)$ *be an indecomposable Weyl family of colligations and suppose* $\operatorname{Ker} A \neq 0$. *Then* $A = 0$ *and the following two cases are possible:*

1) $X(p) = \left(-\sigma_0 p_0 + \sum_{\alpha=1}^{3} \sigma_\alpha p_\alpha,\ H_{\tilde{U},U},\ \varphi,\ E_V\right)$, $\tilde{U}_g = c_g^{*-1}$, $U_g = c_g$;

2) $X(p) = \left(-\tilde{\sigma}_0 p_0 + \sum_{\alpha=1}^{3} \tilde{\sigma}_\alpha p_\alpha,\ H_{\tilde{U},U},\ \varphi,\ E_V\right)$, $\tilde{U}_g = c_g$, $U_g = c_g^{*-1}$

Furthermore, in both cases the channel subspace is either equal to zero or coincident with all of H.

Suppose now A is a nonsingular operator. From (10.9) it follows that in this case the representations $g \to \tilde{U}_g$ and $g \to U_g$ are equivalent. This, clearly, can occur only when $n = m$. But then these representations are unitarily equivalent:

$$\tilde{U}_g = V U_g V^{-1}, \quad V^* = V^{-1}; \tag{10.20}$$

here the matrix of the operator V relative to the canonical basis has the form

$$\left(\begin{array}{c:c} 0 & \underbrace{\begin{matrix} I & & \\ & \ddots & \\ & & I \end{matrix}}_{n} \\ \hdashline \underbrace{\begin{matrix} I & & \\ & \ddots & \\ & & I \end{matrix}}_{n} & 0 \end{array}\right)$$

The results obtained in Chapter IX, § 4, imply that in the present case $A = \lambda B$, $B^2 = I$ and the operator B does not share a common nontrivial invariant subspace with the representation $g \to U_g$. From (10.9) and (10.20) we get

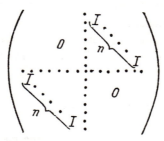

$U_gV^{-1}A = V^{-1}AU_g$, i.e.

$$U_gC = CU_g, \quad C = V^{-1}B. \tag{10.21}$$

Since $B = VC$, $B = B^{-1}$, it follows that $VC = C^{-1}V^{-1}$. Consequently,

$$C = VC^{-1}V^{-1}. \tag{10.22}$$

We now represent H in the form $H = H_1 \oplus H_2$, where H_1 is the invariant under the U_g subspace on which the representation $g \to U_g$ is an n-multiple orthogonal sum of representations $g \to c_g$. Clearly, on H_2 the representation $g \to U_g$ splits into an n-multiple orthogonal sum of representations $g \to c_g^{*-1}$. Moreover,

$$VH_1 = H_2, \; VH_2 = H_1. \tag{10.23}$$

Let P_i be the orthogonal projection onto H_i $(i = 1, 2)$. From (10.21) we see that H_i is invariant under C. Thus $C = C_1P_1 + C_2P_2$, where C_i is the restriction of C to H_i $(i = 1, 2)$. From (10.22) we get $C_2 = VC_1^{-1}V^{-1}$, i.e.

$$C = C_1P_1 + VC_1^{-1}V^{-1}P_2. \tag{10.24}$$

Let μ be an eigenvalue of C_1 and let $\tilde{M}_\mu$ be the corresponding eigenspace. From (10.21) it follows that $\tilde{M}_\mu$ is invariant under the U_g. Let M_μ denote an invariant under the U_g subspace of M_μ on which the representation $g \to U_g$ is irreducible. Clearly, $\dim M_\mu = 2$ and the representation $g \to U_g$ is equivalent on M_μ to the representation $g \to c_g$. From (10.20) we see that VM_μ is invariant under the $\tilde{U}_g$, and since the representations $g \to \tilde{U}_g$ and $g \to U_g$ have the same invariant subspaces, VM_μ is also invariant under the U_g. Further, since $C_2 = VC_1^{-1}V^{-1}$, it follows that VM_μ is invariant under C_2 and

$$C_2h = \frac{1}{\mu} h, \; h \in VM_\mu$$

We therefore denote VM_μ by $M_{\frac{1}{\bar{\mu}}}$. Thus the subspace $M_\mu \oplus M_{\frac{1}{\bar{\mu}}}$ is invariant under C, V and the U_g and hence also under $B = VC$. Since the operator B

and the representation $g \to U_g$ do not have any common nontrivial invariant subspaces,

$$H = M_\mu \oplus M_{\frac{1}{\mu}},$$

i.e. $M_\mu = H_1$, $M_{\frac{1}{\mu}} = H_2$. This means that $n = m = 1$. Consequently, relative to the canonical basis

$$C = \begin{pmatrix} \mu & I & 0 \\ 0 & \frac{1}{\mu} & I \end{pmatrix}, \qquad V = \begin{pmatrix} 0 & I \\ I & 0 \end{pmatrix},$$

$$B = \begin{pmatrix} 0 & I \\ I & 0 \end{pmatrix}\begin{pmatrix} \mu & I & 0 \\ 0 & \frac{1}{\mu} & I \end{pmatrix} = \begin{pmatrix} 0 & \frac{1}{\mu} & I \\ \mu & I & 0 \end{pmatrix}, \quad A = \lambda B = \lambda \begin{pmatrix} 0 & \frac{1}{\mu} & I \\ \mu & I & 0 \end{pmatrix},$$

$$U_g = \begin{pmatrix} c_g^{*-1} & 0 \\ 0 & c_g \end{pmatrix}, \quad U_g = \begin{pmatrix} c_g & 0 \\ 0 & c_g^{*-1} \end{pmatrix}.$$

We note that the admissible subspaces in the present case are 0, M_μ, $M_{\frac{1}{\mu}}$ and H. It is easily verified that either $A - A^* = 0$ or $A - A^*$ is a nonsingular operator. Hence, if $A - A^* = 0$, the channel subspace can be 0, M_μ, $M_{\frac{1}{\mu}}$ or H; while if $A - A^*$ is a nonsingular operator, the channel subspace coincides with all of H.

We have thus proved

Theorem 10.4. *Let* $X(p) = (-N_0 p_0 + \sum_{\alpha=1}^{3} N_\alpha p_\alpha + A, H_{\tilde{U}.U}, \varphi, E_V)$ *be an indecomposable Weyl family of colligations the basic operator of which is nonsingular. Then, relative to the canonical basis of H,*

$$N_k = \begin{pmatrix} \sigma_k & 0 \\ 0 & \tilde{\sigma}_k \end{pmatrix} \ (k = 0, 1, 2, 3), \quad A = \lambda \begin{pmatrix} 0 & \frac{1}{\mu} & I \\ \mu & I & 0 \end{pmatrix}$$

$$\tilde{U}_g = \begin{pmatrix} c_g^{*-1} & 0 \\ 0 & c_g \end{pmatrix}, \quad U_g = \begin{pmatrix} c_g & 0 \\ 0 & c_g^{*-1} \end{pmatrix}.$$

In addition, if $A - A^* = 0$, *the channel subspace can be* 0, M_μ, $M_{\frac{1}{\mu}}$ *or* H; *otherwise* $A - A^*$ *is nonsingular and the channel subspace coincides with all of H.*

An effective method of decomposing arbitrary Weyl families of colligations into a chain of indecomposable Weyl families of colligations is given in [13²].

§4. THE CHARACTERISTIC OPERATOR FUNCTION OF A WEYL FAMILY OF COLLIGATIONS

Let $X(p) = (-N_0 p_0 + \sum_{\alpha=1}^{3} N_\alpha p_\alpha + A,\ H_{\tilde{U},U},\ \varphi,\ E_V)$ be a Weyl family of colligations. Then $(A,\ H_{\tilde{U},U},\ \varphi,\ E_V)$ is a colligation that is invariant with respect to the proper Lorentz group and whose internal representations satisfy the condition $\tilde{U}_g^* = U_g^{-1}$. Making use of the remarks to Theorem 9.1.1, we note that the external representation $g \to V_g$ of a Weyl family of colligations can be assumed to be a unitary representation that satisfies the following two conditions.

α) The space E decomposes into an orthogonal sum of subspaces E_0^c and E_0: $E = E_0^c \oplus E_0$, such that E_0^c reduces the operators V_g. In addition, on E_0^c the representation $g \to V_g$ splits into a direct sum of irreducible representations of the proper Lorentz group that are respectively defined by either the pair $\left(\frac{1}{2},\ \frac{3}{2}\right)$ or the pair $\left(\frac{1}{2},\ -\frac{3}{2}\right)$, while on E_0 one has $V_g = I$.

β) The subspace E_0^c reduces the c.o.f. $S(p)$ of the family of colligations $X(p)$ and, in this connection,

$$S(p) f = f,\ f \in E_0.$$

It will be assumed everywhere below that the external representation $g \to V_g$ of a Weyl family of colligations is unitary and satisfies the above conditions.

We represent the subspaces E_0^c and E_0 in the forms

$$E_0^c = E_+^{(1)} \oplus E_-^{(1)},\ E_0 = E_+^{(2)} \oplus E_-^{(2)},$$

where for any vectors $f_+^{(i)} \in E_+^{(i)}$, $f_-^{(i)} \in E_-^{(i)}$ $(i = 1,\ 2)$ we have $(f_+^{(i)},\ f_+^{(i)}) > 0$, $(f_-^{(i)},\ f_-^{(i)}) < 0$. We define a positive definite metric in E by putting for any $f,\ \tilde{f} \in E$,

$$[f, \tilde{f}] = (f_+^{(1)}, \tilde{f}_+^{(1)}) - (f_-^{(1)}, \tilde{f}_-^{(1)}) + (f_+^{(2)}, \tilde{f}_+^{(2)}) - (f_-^{(2)}, \tilde{f}_-^{(2)}), \quad (10.25)$$

where

$$f = f_+^{(1)} + f_-^{(1)} + f_+^{(2)} + f_-^{(2)};$$
$$\tilde{f} = \tilde{f}_+^{(1)} + \tilde{f}_-^{(1)} + \tilde{f}_+^{(2)} + \tilde{f}_-^{(2)};$$
$$f_+^{(i)}, \tilde{f}_+^{(i)} \in E_+^{(i)};\ f_-^{(i)}, \tilde{f}_-^{(i)} \in E_-^{(i)}\ (i = 1,\ 2).$$

The metric $(f, \tilde{f})$ in E is connected with the metric $[f, \tilde{f}]$ in the following way: $(f, f) = [Jf, \tilde{f}]$, where

$$J = P_+^{(1)} - P_-^{(1)} + P_+^{(2)} - P_-^{(2)}, \quad (10.26)$$

in which $P_+^{(i)} (i = 1, 2)$ is the orthogonal projection onto $E_+^{(i)}$ and $P_-^{(i)}$ $(i = 1,\ 2)$ is the orthogonal projection onto $E_-^{(i)}$. From property α) it follows that there exists in E a basis which is orthonormal with respect to the positive-definite metric (10.15) and relative to which the matrices of the operators V_g and J have the form

$V_g =$

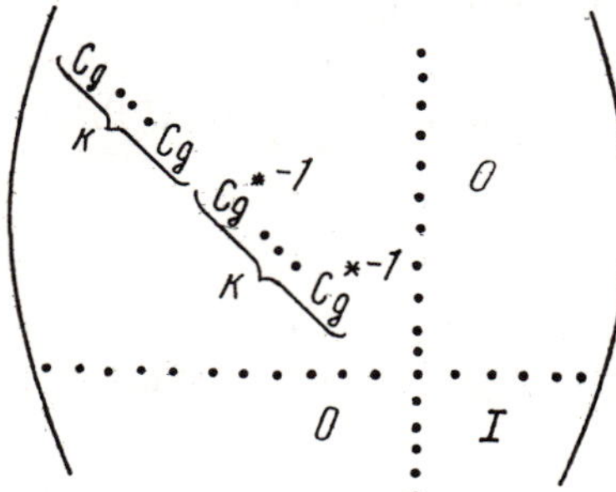

(10.27)

$J =$

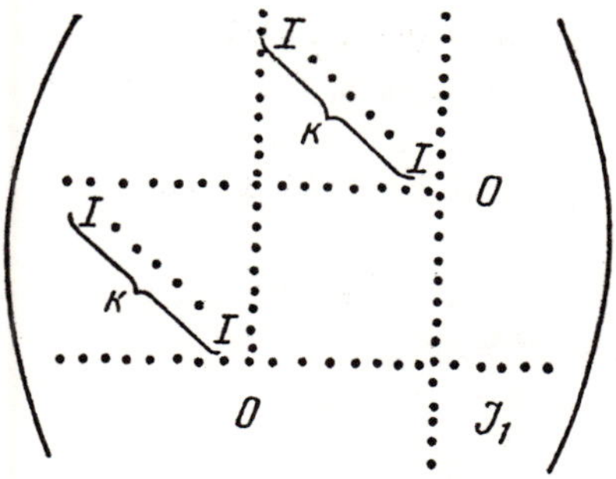

$J_1 = J_1^* = J_1^{-1}$ (10.28)

We will consider below Weyl families of colligations, and hence their c.o.f.'s, on the set D of points $p=(p_0, p_1, p_2, p_3) \in R^4$, satisfying the condition $p_0^2-p_1^2-p_2^2-p_3^2>0$, i.e. the interior of the cone $p_0^2-p_1^2-p_2^2-p_3^2=0$.

Suppose now E is a fixed space with a definite metric, J is an involution ($J = J^* = J^{-1}$) in E and $g \to V_g$ is a representation of the proper Lorentz group on E satisfying condition α) with respect to the metric defined on E by the involution J. Then the operators V_g and J are respectively given by the matrices (10.17) and (10.28) relative to some orthonormal basis of E. We will say that a function $S(p)$ of $p=(p_0, p_1, p_2, p_3) \in R^4$, whose values are linear operators in E, belongs to the $\Omega_{J,V}$ if

(I) it is defined on a domain D_s obtained by removing from the domain D a finite number of surfaces $s^2(p)=\rho_k$ $(k=1, 2, \ldots, l)$, where $s^2(p)=p_0^2-p_1^2-p_2^2-p_3^2$ and the ρ_k are positive numbers;

(II) the subspaces E_0^c and E_0 in condition α) are invariant under $S(p)$ and, in this connection, $S(p)f=f$, $f \in E_0$;

(III) $S(p_0, 0, 0, 0)$ is a rational function of p_0 and, under continuation into the complex plane, belongs to the class Ω_J introduced in §4 of Chapter IV;

(IV) it admits a representation on E_0^c of the form

$$S(p)=\tilde{\omega}(s^2(p))+\sum_{k=0}^{3}\omega_k(s^2(p))p_k, \tag{10.29}$$

in which

$$\left.\begin{aligned}\tilde{\omega}(s^2(p))B_\alpha &= B_\alpha\tilde{\omega}(s^2(p)),\\ \omega_\alpha(s^2(p)) = 2B_\alpha\omega_0(s^2(p)) &= -2\omega_0(s^2(p))B_\alpha\end{aligned}\right\}\ (\alpha=1, 2, 3),$$

where the B_α are the infinitesimal operators of the representation $g \to V_g$ that correspond to hyperbolic rotations in the (p_0, p_α)-plane.

Since the infinitesimal operators B_α satisfy the conditions

$$B_\alpha B_\beta=-B_\beta B_\alpha\ (\alpha, \beta=1, 2, 3;\ \alpha \neq \beta),\ B_\alpha^2=\frac{1}{4}1,$$

in the present case, it is not difficult to see that the product of any two members of $\Omega_{J,V}$ is also a member of $\Omega_{J,V}$.

Let $X(p) = (-N_0p_0 + \sum_{\alpha=1}^{3} N_\alpha p_\alpha + A,\ H_{\tilde{U},U},\ \varphi,\ E_V)$ be a Weyl family of colligations. We introduce the positive definite metric (10.15) in E and let J be the corresponding operator (10.16). We will regard E as a unitary space with metric (10.15) and write the family $X(p)$ in the form $(-N_0p_0 + \sum_{\alpha=1}^{3} N_\alpha p_\alpha + A,\ H_{\tilde{U},U},\ \varphi,\ J,\ E_V)$.

Theorem 10.13. *The c.o.f. $S(p)$ of the Weyl family of colligations $X(p)$ belongs to the class $\Omega_{J,V}$. Conversely, every function $S(p) \in \Omega_{J,V}$ is the c.o.f. of some Weyl family of colligations.*

We will say that a Weyl family of colligations $X(p)$ is *simple* if the linear span of the vectors $A^n\varphi^+f$ $(n = 0,\ 1,\ \ldots;\ f \in E)$ coincides with all of H. Further, we will say that two Weyl families of colligations $X_1(p)$ and $X_2(p)$ with the same E_V and J are *unitarily equivalent* if there exists an isometric mapping V of H_1 onto H_2 such that

$$VN_k^{(1)} = N_k^{(2)}V \quad (k = 0,\ 1,\ 2,\ 3),\quad VA_1 = A_2V,$$
$$V\tilde{U}_g^{(1)} = \tilde{U}_g^{(2)}V,\quad VU_g^{(1)} = U_g^{(2)}V.$$

Clearly, the c.o.f.'s of unitarily equivalent Weyl families of colligations coincide. On the other hand, we have

Theorem 10.14. *Let $X_1(p)$ and $X_2(p)$ be simple Weyl families of operator colligations with the same E_V and J and let $S_i(p)$ $(i = 1,\ 2)$ be the c.o.f. of the Weyl family of colligations $X_i(p)$. If $S_1(p) = S_2(p)$ for all p such that $s^2(p) > \rho$ for some $\rho > 0$ the families of colligations $X_1(p)$ and $X_2(p)$ are unitarily equivalent.*

BIBLIOGRAPHY

1. Ahiezer, N. I., and I. M. Glazman, *The theory of linear operators in Hilbert space,* 2nd rev. ed., Nauka, Moscow, 1966; English transl. of 1st ed., Ungar, New York, 1961. MR **34**, No. 6527.
2. Balakrishnan, A. V.
   1) *On the state space theory of linear systems,* J. Math. Anal. Appl. **14** (1966), pp. 371–391. MR **33**, No. 2463.
   2) *Linear systems with infinite dimensional state spaces,* Proc. Sympos. System Theory (New York, 1965), Polytechnic Press, Polytechnic Inst. Brooklyn, Brooklyn, N.Y., 1965, pp. 69–88. MR **41**, No. 1398.
3. de Branges, L., and J. Rovnyak, *Square summable power series,* Holt, Rinehart and Winston, New York, 1966. MR **35**, No. 5909.
4. Brodskii, M. S. *Triangular and Jordan representations of linear operators,* Nauka, Moscow, 1969; English transl., Transl. Math. Monographs, vol. 32, Amer. Math. Soc., Providence, R.I., 1971. MR **41**, No. 4283.
5. Brodskii, M. S., and M. S. Livsic, *Spectral analysis of non-self-adjoint operators and intermediate systems,* Uspehi Mat. Nauk **13** (1958), no. 1 (79), pp. 3–85; English transl., Amer. Math. Soc. Transl. (2) **13** (1960) MR **20**, No. 7221; **22**, No. 3982, pp. 265–346.
6. Brodskii, V. M.
   1) *Certain theorems on colligations and their characteristic functions,* Funkcional. Anal. i Prolozen. **4** (1970), no. 3, pp. 95–96; Functional Anal. Appl. 4 (1970), MR **42**, No. 3605, pp. 250–251.

2) *Multiplication and division theorems for characteristic functions of an invertible operator,* Acta Sci. Math. (Szeged) **32** (1971), pp. 165–176. (Russian)

7. Brodskii, V. M., I. C. Gohberg, and M. G. Krein
   1) *Definition and basic properties of the characteristic function of a U-node,* Funkcional. Anal. i Prilozen. **4** (1970), no. 1, 88–90; Functional Anal. Appl. 4 (1970), pp. 78–80.
   2) *On the characteristic functions of an invertible operator,* Acta Sci. Math. (Szeged) **32** (1971), pp. 141–164. (Russian)
8. Cekanovskii, E. R. *Generalized extensions of nonsymmetric operators,* Mat. Sb. **68 (110)** (1965), pp. 527–548; English transl., Amer. Math. Soc. Transl. (2) **62** (1967), MR **33**, No. 568, pp. 263–284.
9. Cramer, H. *On the structure of purely nondeterministic stochastic processes,* Ark. Mat. **4** (1961), MR **25**, No. 2631, pp. 249–266.
10. Curtis, C. W. and I. Reiner, *Representation theory of finite groups and associative algebras,* Pure and Appl. Math., vol. 11, Interscience, New York, 1962; Russian transl., Nauka, Moscow, 1969. MR **26**, No. 2519; **40**, No. 1490.
11. Daleckii, Ju. L. and Krein, M. G. *Stability of solutions of differential equations in Banach space*, Nauka, Moscow, 1970; English transl., Transl. Math. Monographs, vol. 43, Amer. Math. Soc., Providence, R.I., 1974.
12. Do Hong Tan, *Equivalent operator colligations,* Teor. Funkcii Funkcional. Anal. i Prilozen. Vyp. 7 (1968), pp. 6–12; (Russian) MR **43**, No. 3833.
13. Dubovoi, V. K.,
   1) *Invariant operator colligations,* Vestnik Har'kov. Gos. Univ. no. 67, Mat.–Meh. Vyp. 36 (1971), pp. 36–71; (Russian) MR **45**, No. 5795.
   2) *Weyl families of operator colligations and the open fields corresponding to them,* Teor. Funkcii Funkcional Anal. i Prilozen. Vyp. 14 (1971), pp. 67–83. (Russian) MR **45**, No. 5796.
14. Gel'fand, I. M., R. A. Minlos, and Z. Ja. Sapiro, *Representations of the rotation group and of the Lorentz group, and their applications,* Fizmatgiz, Moscow, 1958; English transl., Macmillan, New York, 1963. MR **22**, No. 5694.
15. Glazman, I. M., *On expansibility in a system of eigenelements of dissipative operators,* Uspehi Mat. Nauk **13**, (1958), no. 3 (81), 179–181. (Russian) MR **20**, No. 4193.
16. Gohberg, I. C., and M. G. Krein,
   1) *Introduction to the theory of linear nonselfadjoint operators in Hilbert space,* Nauka, Moscow, 1965; English transl., Transl. Math. Monographs, vol. 18, Amer. Math. Soc., Providence, R.I., 1969. MR **36**, No. 3137; **39**, No. 7447.

2) *Theory and applications of Volterra operators in Hilbert space,* Nauka, Moscow, 1967; English transl., Transl. Math. Monogrpahs, vol. 24, Amer. Math. Soc., Providence, R.I., 1970. MR**36**, No. 2007.

17. H. Helson *Lectures on invariant subspaces,* Academic Press, New York, 1964. MR **30**, No. 1409.
18. Ito, K., *Stochastic processes,* Iwanami Shoten, Tokyo, 1957 (Japanese); Russian transl., Chaps. I–III, IL, Moscow, 1960. MR **23**, No. A2917.
19. Jancevic, A. A., *Operator J-colligations and the associated open systems,* Teor. Funkcii Funkcional. Anal. i Prilozen. Vyp. 17 (1972), pp. 215–220. (Russian)
20. Kacnel'son, V. E., *Conditions for a system of root vectors of certain classes of operators to be a basis,* Funkcional. Anal. i Prilozen. **1**, (1967), no. 2, pp. 39–51; Functional Anal. Appl. **1** (1967), MR **35**, No. 4757, pp. 122–132.
21. Karhunen, K., *Uber lineare Methoden an Wahrscheinlichkeitsrechnung,* Ann. Acad. Sci. Fenn. Ser. AI Math.–Phys. No. 37 (1947). MR **9**, pp. 292.
22. Kircev, K. N., *A certain class of nonstationary random processes,* Teor. Funkcii Funkcional. Anal. i Prilozen. Vyp. 14 (1971), pp. 150–169; (Russian) MR **45**, No. 7793.
23. Kolmogorov, A. N., *Grundbegriffe der Wahrscheinlichkeitsrechnung,* Springer, Berlin, 1933; Russian transl., ONTI, Moscow, 1936; English transl., Chelsea, New York, 1950. MR **11**, p. 374.
24. Kuzel', A. V.,
    1) *The reduction of unbounded non-selfadjoint operators to triangular form,* Dokl. Akad. Nauk SSSR **119** (1958), pp. 868–871; (Russian) MR **20**, No. 6041.
    2) *Spectral analysis of unbounded non-selfadjoint operators,* Dokl. Akad. Nauk SSSR **125** (1959), pp. 35–37; (Russian) MR **22**, No. 2905.
    3) *The spectral analysis of quasi-unitary operators in a space with indefinite metric,* Teor. Funkcii Funkcional. Anal. i Prilozen. Vyp. 4 (1967), pp. 3–27; (Russian) MR **36**, No. 5736.
    4) *Spectral analysis of unbounded nonselfadjoint operators in a space with indefinite metric,* Dokl. Akad. Nauk SSSR **178** (1968), pp. 31–33; Soviet Math. Dokl. **9** (1968), pp. 25–27. MR **36**, No. 5737.
25. Lang, S., *Algebra,* Addison–Wesley, Reading, Mass., 1965; Russian transl., Mir, Moscow, 1968. MR **33**, No. 5416.
26. Lévy, P., *Processus stochastiques et mouvement Brownier,* Suivi d'une note de M. Loève, Gauthier–Villars, Paris, 1948. MR **10**, No. 551.
27. Livsic, M. S.,
    1) *Operators, oscillations, waves. Open systems,* Nauka, Moscow, 1966;

English transl., Transl. Math. Monographs, vol. 34, Amer. Math. Soc., Providence, R.I., 1973. MR **38**, No. 1922.

2) *Nonunitary representations of groups,* Funkcional. Anal. i Prilozen. **3** (1969), no. 1, pp. 62–70; Functional Anal. Appl. **3** (1969). pp. 51–57. MR **39**, No. 7464.

28. Livsic, M. S., and V. P. Potapov, *A theorem on the multiplication of characteristic matrix functions,* Dokl. Akad. Nauk SSSR **72** (1950), pp. 625–628. (Russian) MR **11**, No. 669.
29. Markus, A. S., *Expansion in root vectors of a slightly perturbed self-adjoint operator,* Dokl. Akad. Nauk SSSR **142** (1962), pp. 538–541; Soviet Math. Dokl. **3** (1962), MR **27**, No. 1837, pp. 104–107.
30. Mesarović, M. D., Editor, *Views on general systems theory,* Proc. Second Systems Sympos. (Case Inst. Tech., 1963), Wiley, New York, 1964; Russian transl. of selected articles, *General systems theory,* Mir, Moscow, 1966. MR **34**, No. 8861; **37**, No. 1196.
31. Mukminov, B. R., *On expansion with respect to the eigenfunctions of dissipative kernels,* Dokl. Akad. Nauk SSSR **99** (1954), pp. 499–502; (Russian) MR**16**, No. 830.
32. Naimark, M. A., *Linear representations of the Lorentz group*, Fizmatgiz, Moscow, 1958; English transl., Macmillan, New York, 1964. MR **21**, No. 4995; **30**, No. 1211.
33. Nikol'skii, N. K., and B. S. Pavlov, *Eigenvector bases of completely nonunitary contractions and the characteristic function,* Izv. Akad. Nauk SSSR Ser. Mat. **34** (1970), pp. 90–133; Math. USSR Izv. **4** (1970), MR **41**, No. 9027, pp. 91–134.
34. Polyjackii, V. T. *On the reduction of quasi-unitary operators to a triangular form,* Dokl. Akad. Nauk SSSR **113** (1957), pp. 756–759; (Russian) MR **19**, No. 873.
35. Rasevskii, P. K., *Riemannian geometry and tensor analysis,* GITTL, Moscow, 1953; German transl., Hochschulbücher für Mathematik, Band 42, VEB Deutscher Verlag, Berlin, 1959. MR **16**, No. 1051; **21**, No. 2258.
36. Rozanov, Ju. A., *Stationary random processes,* Fizmatgiz, Moscow, 1963; English transl., Holden–Day, San Francisco, Calif., 1967. MR **28**, No. 2580; **35**, No. 4985.
37. Sahnovic, L. A., *Dissipative operators with absolutely continuous spectrum,* Trudy Moskov. Mat. Obsc. **19**, (1968), pp. 211–270; Trans. Moscow Math. Soc. **19** (1968), MR **40**, No. 3350, pp. 233–298.
38. Straus, A. V., *Characteristic functions of linear operators,* Izv. Akad. Nauk SSSR Ser. Mat. **24** (1960), pp. 43–74; English transl., Amer. Math. Soc. Transl. (2) **40** (1964), MR **25**, No. 4363, pp. 1–37.

39. Svarcman, Ja. S., *A functional model of a completely continuous dissipative assembly,* Mat. Issled. **3** (1968), no. 3 (9), pp. 126–138. (Russian) MR **41**, No. 2441.

40. Sz.-Nagy, B. and Foias, C., *Analyse harmonique des opérateurs de l'espace de Hilbert,* Masson, Paris; Akad. Kiadó, Budapest, 1967; English rev. transl., North-Holland, Amsterdam; American Elsevier, New York; Akad. Kiadó, Budapest, 1970; Russian transl., Mir, Moscow, 1970. MR **37**, No. 778; **43**, No. 947; No. 948.

41. Vaksman, L. L., *On characteristic operator-functions of Lie algebras,* Vestnik Har'kov. Univ. no. 83, Mat.–Meh. Vyp. 37 (1972), pp. 41–45. (Russian)

42. Zadeh, L. A. and Desoer, C. A., *Linear system theory,* McGraw–Hill, New York, 1963.

43. Zmud', E. M., *Operator colligations and Witt groups,* Teor. Funkcii Funkcional. Anal. i Prilozen. Vyp. 13 (1970), (Russian)

INDEX